Advances in Intelligent Systems and Computing

Volume 718

Series editor

Janusz Kacprzyk, Polish Academy of Sciences, Warsaw, Poland
e-mail: kacprzyk@ibspan.waw.pl

The series “Advances in Intelligent Systems and Computing” contains publications on theory, applications, and design methods of Intelligent Systems and Intelligent Computing. Virtually all disciplines such as engineering, natural sciences, computer and information science, ICT, economics, business, e-commerce, environment, healthcare, life science are covered. The list of topics spans all the areas of modern intelligent systems and computing.

The publications within “Advances in Intelligent Systems and Computing” are primarily textbooks and proceedings of important conferences, symposia and congresses. They cover significant recent developments in the field, both of a foundational and applicable character. An important characteristic feature of the series is the short publication time and world-wide distribution. This permits a rapid and broad dissemination of research results.

More information about this series at http://www.springer.com/series/11156

Anurag Mishra · Anirban Basu
Vipin Tyagi
Editors

Silicon Photonics & High Performance Computing

Proceedings of CSI 2015

Springer

Editors
Anurag Mishra
Deen Dayal Upadhyaya College
University of Delhi
New Delhi, Delhi
India

Vipin Tyagi
Department of Computer Science and Engineering
Jaypee University of Engineering and Technology
Guna, Madhya Pradesh
India

Anirban Basu
Department of Computer Science and Engineering
Visvesvaraya Technological University
Belgaum, Karnataka
India

ISSN 2194-5357 ISSN 2194-5365 (electronic)
Advances in Intelligent Systems and Computing
ISBN 978-981-10-7655-8 ISBN 978-981-10-7656-5 (eBook)
https://doi.org/10.1007/978-981-10-7656-5

Library of Congress Control Number: 2017961500

Printed on acid-free paper

This Springer imprint is published by Springer Nature
The registered company is Springer Nature Singapore Pte Ltd.
The registered company address is: 152 Beach Road, #21-01/04 Gateway East, Singapore 189721, Singapore

Preface

The last decade has witnessed remarkable changes in IT industry, virtually in all domains. The 50th Annual Convention, CSI-2015, on the theme “Digital Life” was organized as a part of CSI@50, by CSI at Delhi, the national capital of the country, during December 2–5, 2015. Its concept was formed with an objective to keep ICT community abreast of emerging paradigms in the areas of computing technologies and more importantly looking at its impact on the society.

Information and Communication Technology (ICT) comprises of three main components: infrastructure, services, and product. These components include the Internet, Infrastructure-based/infrastructure-less wireless networks, mobile terminals, and other communication mediums. ICT is gaining popularity due to rapid growth in communication capabilities for real-time-based applications. The Silicon Photonics & High Performance Computing includes design and analysis of parallel and distributed systems, embedded systems, and their applications in scientific, engineering, and commercial deployment. CSI-2015 attracted over 1500 papers from researchers and practitioners from academia, industry, and government agencies, from all over the world, thereby making the job of the Programme Committee extremely difficult. After a series of tough review exercises by a team of over 700 experts, 565 papers were accepted for presentation in CSI-2015 during the 3 days of the convention under ten parallel tracks. The Programme Committee, in consultation with Springer, the world’s largest publisher of scientific documents, decided to publish the proceedings of the presented papers, after the convention, in ten topical volumes, under ASIC series of the Springer, as detailed hereunder:

1. Volume # 1: ICT Based Innovations
2. Volume # 2: Next Generation Networks
3. Volume # 3: Nature Inspired Computing
4. Volume # 4: Speech and Language Processing for Human-Machine Communications
5. Volume # 5: Sensors and Image Processing
6. Volume # 6: Big Data Analytics

7. Volume # 7: Systems and Architecture
8. Volume # 8: Cyber Security
9. Volume # 9: Software Engineering
10. Volume # 10: Silicon Photonics & High Performance Computing.

We are pleased to present before you the proceedings of Volume # 10 on "Silicon Photonics & High Performance Computing." Presently, the data is growing exponentially. This data is an outcome of continuous research and development, demanding our serious concerns toward its safety. Computing is all about processing data in a meaningful manner; hence, it assumes a significant space in today's research arena. The present CSI-2015 track, Silicon Photonics and High Performance Computing, is even more relevant due to this specific reason. It is our pleasure and honor to serve as the editor of the proceeding of this track. This is a constant and consistent activity organized and conducted by the Computer Society of India, and it is a matter of satisfaction that the Springer has agreed to publish all its proceedings.

This volume is unique in its coverage. It has received papers from all research domains—from photonics/optical fiber communication systems used for different applications to high-performance computing and cloud computing for social media analytics and other very relevant high-ended applications such as supply chain analysis and underwater signal processing. The articles submitted and published in this volume are of sufficient scientific interest and help to advance the fundamental understanding of ongoing research, applied or theoretical, for a general computer science audience. The treatment of each topic is in-depth, the emphasis is on clarity and originality of presentation, and each paper is adding insight into the topic under consideration. We are hopeful that that this book will be an indispensable help to a broad array of readers ranging from researchers to developers and will also give significant contribution toward professionals, teachers, and students.

A great deal of effort has been made to realize this book. We are very thankful to the team of Springer who have constantly engaged us and others in this process and have made the publication of this book a success. We are sure this engagement shall continue in future as well and both Computer Society of India and Springer will choose to collaborate academically for the betterment of the society at large. Under the CSI-2015 umbrella, we received over 100 papers for this volume, out of which 15 papers are being published, after rigorous review processes, carried out in multiple cycles.

On behalf of organizing team, it is a matter of great pleasure that CSI-2015 has received an overwhelming response from various professionals from across the country. The organizers of CSI-2015 are thankful to the members of *Advisory Committee, Programme Committee, and Organizing Committee* for their all-round guidance, encouragement, and continuous support. We express our sincere gratitude to the learned *Keynote Speakers* for support and help extended to make this event a grand success. Our sincere thanks are also due to our *Review Committee Members* and the *Editorial Board* for their untiring efforts in reviewing the manuscripts, giving suggestions and valuable inputs for shaping this volume.

We hope that all the participated delegates will be benefitted academically and wish them for their future endeavors.

We also take the opportunity to thank the entire team from Springer, who have worked tirelessly and made the publication of the volume a reality. Last but not least, we thank the team from Bharati Vidyapeeth's Institute of Computer Applications and Management (BVICAM), New Delhi, for their untiring support, without which the compilation of this huge volume would not have been possible.

New Delhi, India
Belgaum, India
Guna, India
March 2017

Anurag Mishra
Anirban Basu
Vipin Tyagi

The Organization of CSI-2015

Chief Patron

Padmashree Dr. R. Chidambaram, Principal Scientific Advisor, Government of India

Patrons

Prof. S. V. Raghavan, Department of Computer Science, IIT Madras, Chennai
Prof. Ashutosh Sharma, Secretary, Department of Science and Technology, Ministry of Science of Technology, Government of India

Chair, Programme Committee

Prof. K. K. Aggarwal
Founder Vice Chancellor, GGSIP University, New Delhi

Secretary, Programme Committee

Prof. M. N. Hoda
Director, Bharati Vidyapeeth's Institute of Computer Applications and Management (BVICAM), New Delhi

Advisory Committee

Padma Bhushan Dr. F. C. Kohli, Co-Founder, TCS
Mr. Ravindra Nath, CMD, National Small Industries Corporation, New Delhi
Dr. Omkar Rai, Director General, Software Technological Parks of India (STPI), New Delhi

Editorial Board

Contents

About the Editors

Dr. Anurag Mishra has more than 23 years of research and teaching experience. He is presently working as Associate Professor of Electronics, Deen Dayal Upadhyaya College, University of Delhi, India. He is actively involved in research in information security and digital watermarking of images and video in particular, intelligent systems employed for image processing using soft computing techniques such as artificial neural networks, fuzzy systems, support vector machines, and extreme learning machines. He has developed fuzzy inference system-based models for coded image transmission over wireless channels. Additionally, he also extensively uses hybrid techniques such as neuro-fuzzy systems and GA-BPN systems for different image processing applications.

Dr. Anirban Basu holds a master's and Ph.D. in Computer Science and has more than 35 years of experience in academia, advanced research and development, commercial software industry, consultancy, and corporate training. He has worked at the Indian Statistical Institute, Kolkata, in 1979 and joined CDAC, Pune, India, in 1989 to play a key role in the development of India's first supercomputer PARAM. He has also worked at Siemens Information Systems Ltd and Computer Associates TCG in senior/top management positions. He has published 80 research papers in respective national and international journals and authored 6 books including one on "Software Quality Assurance, Testing and Metrics" published by PHI.

Dr. Vipin Tyagi works at the Department of Computer Science and Engineering, Jaypee University of Engineering and Technology, Raghogarh, Guna (MP), India, and is the Regional Vice President of Region 3 of the Computer Society of India (CSI). He is also associated with the CSI Special Interest Group on Cyber Forensics. He has over 20 years of teaching and research experience. He was the President (Engineering Sciences) of the Indian Science Congress Association for the term 2010–2011. He is a Life Fellow of the IETE, New Delhi, India.

Tackling Supply Chain Management Through High-Performance Computing: Opportunities and Challenges

Prashant R. Nair and S. P. Anbuudayasankar

Abstract Conversion of a supply chain to value chain requires agility, adaptability, communication, collaboration, decision support, elasticity, robustness, sensitivity, and visibility. High-Performance Computing (HPC) and cloud computing systems bring additional benefits of scalability, integration, portability, processing power, storage, and interoperability for Supply Chain Management (SCM). Mega corporations and retail giants like Wal-Mart and Pratt & Whitney have deployed HPC for SCM and thereby achieved efficient and effective data administration and analysis. But the usages of HPC for SCM are restricted to a few large enterprises. Deployment challenges include difficulty in migrating from legacy to high-end systems, high cost, lack of skilled manpower, and application software. HPC along with the cloud computing paradigm integrated in the Social Mobile Analytics and Cloud (SMAC) stack can emerge as a game changer to integrate all stakeholders in the value chain into a social network real-time and actionable intelligence. The access to updates and timely information from all supply chain partners will also transform enterprises to be forecast-driven as opposed to their conventional demand-driven nature.

Keywords Supply chain management (SCM) · High-Performance Computing (HPC) · Cloud computing · SMAC

P. R. Nair (✉)
Department of Computer Science Engineering,
Amrita School of Engineering, Amrita Vishwa Vidyapeetham University,
Amrita Nagar, Coimbatore 641112, India
e-mail: prashant@amrita.edu

S. P. Anbuudayasankar
Department of Mechanical Engineering,
Amrita School of Engineering, Amrita Vishwa Vidyapeetham University,
Amrita Nagar, Coimbatore 641112, India

A. Mishra et al. (eds.), *Silicon Photonics & High Performance Computing*,
Advances in Intelligent Systems and Computing 718,
https://doi.org/10.1007/978-981-10-7656-5_1

1 Introduction

Conversion of a supply chain to value chain requires agility, adaptability, communication, collaboration, decision support, elasticity, robustness, sensitivity, and visibility. Enterprises are now grappled with greater competition due to the influence of Internet, social media, and information superhighway. This is further complicated by pricing pressures, outsourcing, and globalization. This has resulted in enterprises increasingly offshoring both service and manufacturing bases to low-cost and emerging geographies and economies.

ICT tools enable all processes of supply chain planning and execution. Deployment of ICT across the supply chain has become the reason behind the competitive edge for many enterprises [1]. Widespread adoption of technologies like ERP, RFID [2], intelligent agents [3], transportation systems, barcodes, and inventory control systems has brought about the outcome of better transparency, visibility, and communication at both intra- and inter-enterprise levels. This will in turn bring about better resilience, adaptability, responsiveness, and decision support and thereby help enterprises gain a competitive edge. Actionable intelligence and seamless information will be made available on demand and real time with the additional provision of analytics. One prime barrier is not having uniformity in the legacy information and transaction processing systems of suppliers and stakeholders in the supply chain. A lion share of data needed by enterprises for their supply chains is vested with suppliers, transporters, and warehouses, who have their distinct information technology and systems and proprietary ERP or supply chain software solutions. Another relevant issue is that enterprises are groping in the dark to get an integrated image of their inventory, operations, and work flow, as majority of them use legacy systems, which were usually intended for single branch and not across a network of branches, divisions, suppliers, and partners [4]. This paper provides an overview of the technologies of High-Performance Computing (HPC) and cloud computing systems and its usage, application areas, and challenges in the context of deployment in Supply Chain Management (SCM) settings.

HPC and cloud systems for SCM bring additional benefits of scalability, integration, portability, processing power, storage, and interoperability for SCM. Mega corporations and retail giants like Wal-Mart and Pratt & Whitney have deployed HPC for SCM and thereby achieved efficient and effective data administration and analysis.

Integration of disruptive technologies like Social, Mobile, and Analytics along with Cloud to form SMAC stack is poised to be the next wave of enterprise computing. By 2020, IDC estimates that corporate spending on buying or building IT/IS solutions worldwide could be north of the US$5 trillion mark. IDC estimates that 80% of the revenue will be motivated by the SMAC stack, which is the seamless combo of these SMAC technologies [5]. Technologies within SMAC complement and supplement each other and collectively bring a force multiplier effect to transform supply chains into value chains. The resultant value chain would boast of the advantages of resilience, agility, collaboration, scalability, and visibility [6]. This would benefit enterprises across all verticals and geographies.

2 High-Performance Computing (HPC)

HPC also referred to as supercomputing in the more colloquial sense started emerging more than five decades back with systems having parallel processors working in tandem. Distributed computing has come of age in the present day. There are several massively parallel supercomputers equipped with several hundred thousand high-end CPUs. Since 2013, Tianhe-2 which in Mandarin means Milky Way-2, 33.86 petaflop supercomputer developed at National University of Defense Technology, based at Guangzhou, China is currently numero uno in the Top 500 list of high-performance computers. Price ranges of HPC systems start from US$10,000 to hundreds of millions of dollars each.

Supercomputing is a unique kind of distributed computing, wherein processors or cores in individual processors synchronize their working and actions. A higher form of distributed computing is cluster computing, when these systems operate together as though they are one system. Typical HPC high-end computational science and engineering problems in molecular modeling, weather forecasting, drug discovery, computational fluid dynamics (CFD), CAD/CAM, 3D printing, rapid prototyping, and simulation are handled through super computing. Cray leads the market space followed by IBM, Intel, and SGI among the prime movers. Dana Holding Corporation provides supercomputing infrastructure for US department of defense (DoD) for its supply chain operations. HPCs are also an important building block of the "cloud" by being the backbone for data centers and server farms who host services on the cloud. Provision for big data analytics on the cloud infrastructure adds to the sheen. Efficiency is further enhanced by virtualization technology, which creates virtual machines through hypervisors on the cloud. Associated with cloud is the deployment of large server farms for storage. This is very eco-friendly and is often referred to as green computing as firms need not invest on data centers or servers and can easily subscribe to these.

In 2011, Amazon public cloud, Amazon EC2 unveiled Cluster Compute EC2 instance, an early deployment of high-performance systems for the cloud. EC2 consists of two eight-core Xeon processors on a 10b network. In 2010, SGI started on-request cloud HPC service called Cyclone [7]. However, HPC is only accounting for around 3% of the computing industry. HPC promotes innovation for enterprises especially when key business processes are integrated into it.

3 Cloud Computing

Forrester Research estimates the total worth of the cloud and associated technology economy worldwide to be $241 billion by 2020 [8]. Prominent ERP solution providers like Baan and SAP are increasingly selling cloud versions of their products. These thin solutions reduce the Total Cost of Ownership (TCO) for companies and contribute to improving profits as well as sales.

Cloud computing is a type of user-friendly convenience computing, where all components like software, hardware, storage, infrastructure, and platform are subscribed to as per need of the client. Users access various applications through an Internet browser. Complementing this is the proliferation of vast data centers in which all software and databases are resident using virtualization with real-time and on-demand access. Cloud is a carbon-positive technology as it enterprises especially SMEs need not maintain large number of servers, who occupy so much room and guzzle power. Cloud computing utilizes software services shared over a network.

Primary cloud is classified as public and private cloud. Public cloud services are on demand through the public internetwork. Amazon, Apple, and Google have architected massive public clouds, which are accessed by millions on a daily basis with a host of value-added services. Private cloud is primarily meant for enterprise internal workflow including collaboration with their units in the intranet and with their partners in the extranet. These could also be outsourced to an external agency. Recently, there are rapid developments to construct hybrid clouds that mingle best attributes of private and public clouds. In these, the order of usage is initially private and on complete utilization of this, migrates to public cloud [9].

4 HPC and Cloud Deployments for SCM

Major HPC advantages like scalability, integration, and interoperability are like a wish list in SCM deployments. This renders efficient management and analysis of the big data generated from multifarious sources. A unified view of the supply chain is preferred by all suppliers, partners, customers, logistic providers, and other stakeholders. The world's largest retail giant, Wal-Mart stocks over 500,000 unique products. Empanelled suppliers in their network battle intensely for shelf space and SKU. Retailers continually need visibility of their inventory from point of sale to warehouse as also the patterns of customer preferences and buying behavior. Many world-class companies in the aerospace, life sciences, and automotive industries are using HPC and mandating their supplier network to plug in.

Wal-Mart is an early adopter of HPC for SCM. HPC is used for the supply chain activities like store, resource, and shelf space planning as also have visibility of all their stores from their headquarters at Bentonville, USA. Pratt and Whitney, the leading aerospace company, simulates inventory positions using supercomputing systems. Various suppliers and partners are also encouraged to plug into the system [10]. However apart from some large enterprises, who have touched the tip of the iceberg, many are yet to leverage the power of HPC or the force multiplier effect through the use of cloud and SMAC technologies. Perhaps, the biggest challenge would be to convince all partners to drift from existing legacy computers to cloud or supercomputers.

Cloud computers and its component services such as Software as a Service (SaaS), Hardware as a Service (HaaS), and Platform as a Service (PaaS) are a dominant force in enterprise computing today. This game-changing technology integrates all stakeholders be it suppliers, transporters, warehouses, customers, distributors, or 3PL providers into a global extranet which encompasses the extended supply chain. The front-end of this resembles a community which can interact like a social network. Typical data points are information on prices, delivery, or production schedules, inventory positions, service options, and updates. The access to updates and timely information from all supply chain partners will also transform enterprises to be forecast-driven as opposed to their conventional demand-driven nature [11]. Major ERP vendors like Oracle, SAP, and Baan are also moving onto the cloud bandwagon by offering cloud-based versions of their product offerings.

Supply chain planning and execution processes which have been migrated to cloud solutions comprise demand planning and forecasting, logistics and e-procurement, distribution channeling, inventory tracking and management, storage and transportation systems. Several vendors like IBM, Virtual Computer Corporation, James Donald Armstrong (JDA) Software, and Ariba are offering HPC and cloud-based services for supply chain processes [12]. Since 2011, FedEx has a private cloud from the service provider, cloudX where sales workflow and customer engagement have been migrated. COSCO Logistics, headquartered at Beijing, is the largest 3PL company of China. This giant also holds the distinction of being world's second biggest marine shipping enterprise. COSCO takes advantage of SaaS service and has mandated their partners to use their logistics management software [13].

Intel was able to cut flab by substituting several of their order clerks using applications on the cloud. Supply chain processes like demand and supply planning and prediction, inventory tracking and replenishing, sourcing, and logistics were migrated to the cloud [14]. However, enterprises who have go for HPC either stand-alone or in conjunction with cloud or SMAC will have to come across hurdles like legacy computers and software solutions among their suppliers and stakeholders as part of their extended supply chain. Yet another challenge is privacy issues in hybrid, private. and public clouds, social media, and smart phones. Lack of skilled manpower and inter-operable application software are some of the other barriers.

JDA cloud-based supply chain solutions can offer adaptive supply chain capabilities for an extended enterprise ecosystem consisting of suppliers, partners, transporters, storage facilities, and distribution networks. An intelligent supply planning optimization workflow produces a master plan that delivers inclusive and exhaustive analysis and visibility, as well as proactively identifies constraints and exceptions. Popular cloud-based features in this package include [15]

- Inventor, transportation, distribution, and capacity planning
- Demand forecasting, shaping, and prioritization
- Exception-based alerts
- What-if analysis

5 Conclusion

HPC solutions, both stand-alone as also in conjunction with cloud or with SMAC stack have started making inroads into various processes and activities for SCM, planning, and execution. These systems help in providing agility, adaptability, communication, collaboration, decision support, elasticity, robustness, sensitivity, scalability, portability, storage, processing power, data integration, interoperability, and visibility to supply chains. HPC along with the cloud computing paradigm integrated in the Social Mobile Analytics and Cloud (SMAC) stack can emerge as a game changer to integrate all stakeholders in the value chain into a social network real-time and actionable intelligence. Cloud computing can join together all supply chain stakeholders into a real-time and interconnected social network like community with real-time updates. SMAC and cloud deployments for SCM are expanding fast with some vendors offering Software as a Service based solutions. However, HPC deployments for SCM are limited to some big players. Deployment challenges include difficulty in migrating from legacy to high-end systems, high cost, lack of skilled manpower, and application software.

References

1. Nair PR, Balasubramaniam OA (2010) IT enabled supply chain management using decision support systems. CSI Comm 34(2):34–40
2. Nair PR (2012) RFID for supply chain management. CSI Comm 36(8):14–18
3. Nair PR (2013) E-supply chain management using software agents. CSI Comm 37(4):13–16
4. Toka A, Eirini A, Antonios A, Konstantinos A (2013) Cloud computing in supply chain management: an overview. E-logistics E-Supply Chain Manage 218–231
5. Frank M (2012) Don't get SMACked: how social, mobile, analytics and cloud technologies are reshaping the enterprise. Cognizant Future of Work
6. Chandrasekharan R, Udhas P (2013) The SMAC code: embracing new technologies for future business.CII and KPMG report information. http://www.kpmg.com/IN/en/IssuesAndInsights/ArticlesPublications/Documents/The-SMAC-code-Embracing-new-technologies-for-future-business.pdf
7. Betts B (2012) HPC cloud: supercomputing to go. E&T 7(2)
8. King J The rise of the Cloud. Supp Chn Europe information http://www.scemagazine.com/the-rise-of-the-cloud/
9. Sujay R (2011) Hybrid cloud: a new era. Int J Comp Sci Tech 2(2):323–326
10. Conway S (2006) HPC and supply chain management. HPC Wire, Sep 15
11. Christopher M (2000) The agile supply chain: competing in volatile markets. Ind Mark Manage 29:37–44
12. McCrea B (2012) Supply chain technology: cloud computing breakthrough. Logistic manage. http://www.logisticsmgmt.com/article/supply_chain_technology_cloud_breakthrough
13. Harris J, Alter A (2010) Cloudrise: rewards and risks at the dawn of cloud computing. Accenture. http://itri.uark.edu/Cloud_Computing_-_Theirs_Mine_and_Ours-v4.pptx

14. Schramm T, Nogueira S, Jones D (2011) Cloud computing and supply chain: a natural fit for the future: Logistic manage. http://www.logisticsmgmt.com/article/cloud_computing_and_supply_chain_a_natural_fit_for_the_future
15. JDA Company information. http://www.jda.com/view/press-release/planning-in-the-cloud-jda-software-reshapes-supply-chain-management-for-manufacturers/

Energy Theft Identification in Smart Grid

K. Govinda, Rishav Shav and Surya Prakash

Abstract Smart grid is a new generation of electrical grid communication with high management of power flow control, self-healing, energy efficiency, and security through the digital communication networks and technologies. To develop a smart grid from the existing power grid, we need to integrate ICT infrastructures with grid and management of grid has to be automated in the smart way, this requires sensing technologies, distributed communication, and pervasive computing frameworks to make the smart grid more efficient and secure. Theft identifying is one of the major issues faced by many service providers and this makes huge loss to the power management and the provider. This paper proposes a secure power management and theft identification in the smart grid.

Keywords Smart grid · Power theft · Infrastructure · Smart household meter
Smart line meter

1 Introduction

Smart grid is the future power grid with the integration of electrical power grid and ICT which is developing all over world, it is a fully sustainable form of reliable and green electrical energy in existing network with advanced technologies and communication devices to manage the system in both sides [1]. These advanced methodology and frameworks provide a great flexibility and management, this also possesses a new class of risk [2]. Smart grid is one of the most critical infras-

K. Govinda (✉) · R. Shav
SCSE, VIT University, Vellore, India
e-mail: kgovinda@vit.ac.in

R. Shav
e-mail: rishav2105@gmail.com

S. Prakash
SITE, VIT University, Vellore, India
e-mail: suryaprakash498@gmail.com

A. Mishra et al. (eds.), *Silicon Photonics & High Performance Computing*,
Advances in Intelligent Systems and Computing 718,
https://doi.org/10.1007/978-981-10-7656-5_2

tructures that are augmented by the large-scale ICT and renewable energy integration [3], even with all the crises, smart grid is the best infrastructure to handle a large set of management system that is distributed in the network. To provide grid monitoring and control capabilities, numerous power applications are necessary to exist [4]. Every year, the utility provider company fares their power theft from 20 to 30% and to that power, ministry loss is more than Rs. 125 billion [7], at this stage, service providers took several steps to manage the distribution system, but this is not enough to handle the power theft. This paper proposes a new model to handle power losses in distribution system.

2 Literature Review

In [1], the author has proposed a security framework using location-based security for protecting the SG infrastructure which is designed based on the algebraic code based cryptosystems, they chose it for smart grids to create location-based security applications.

In [2], the authors have done a study on threads on smart grid and solved it through system engineering and fault management concepts and expanded the range potential, range behaviors, and outcomes in the grid technologies with fault tolerance.

In [3], the authors have investigated challenges and security issues in smart grid, some important issues include privacy issue, identity spoofing, and so on and challenges such as mobility, scalability, deployment, and so on.

In [4], the author has discussed about the cyber security threats in smart grid and focused particularly on government grid infrastructures threats and measures to control and monitor the systems from cyber attacks for smart grid environments.

In [5], the author has done a survey on smart grid cyber security communications which elaborates the threats and vulnerabilities, solution proposed to these problems and relies on the system to make smart grid communication secured.

In [7], the authors have proposed a power theft monitoring system using GSM module and integrated the part used for the project and discussed about it briefly; the components used to implement for the projects such as sensors, circuits, etc., are discussed.

3 Proposed Method

Smart grid technologies are the future power grid that functions with the renewable energy, sensors, and smart meters to provide an efficient power usage and performance management [6, 8, 9]. Energy power resource has the highest priority in every field and thus it makes it as a huge resource and we provide a framework to manage it and identify the theft in the field of line as shown in Fig. 1.

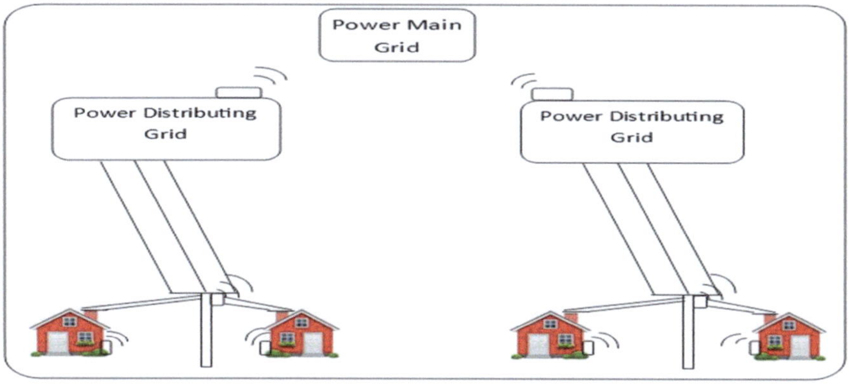

Fig. 1 System architecture

3.1 Smart House Holds Meters (SHHM)

This meter has the GSM module which has unique ID given for the customer and unique GSM number also, which is attached with Arduino board and the board is connected to the meter and communicates with sensors and sends signal from the house to the power distributing grid, the meter sends the power consumption level to power distributing grid in a particular interval of time (for example, every 1 h) and these values are stored in the server and to get processed, these values are indexed in the table on the customer id, every usage based on time is stored separately in another database table. In this way, we know how much power the customer consumes in a particular interval of time and monitor the power activity of customer Fig. 2.

Smart line meter which will be placed in the line post of the house where the connection coming from the line to house is through smart line meter and this meter communicates to the power distributing grid in the same interval of time as smart household meter[10, 11]. This SLM will consist of a number of sockets to get input from the line and output sockets to provide connection to the house, the line number denotes the customer connected to the concerned port which will be stored in the server (for example, 1 means the index will be stored), such that all the sockets readings will be communicated to the Arduino, Arduino will send signal to the GSM module and the power distributing grid will receive it and store it in database.

Fig. 2 Smart meter

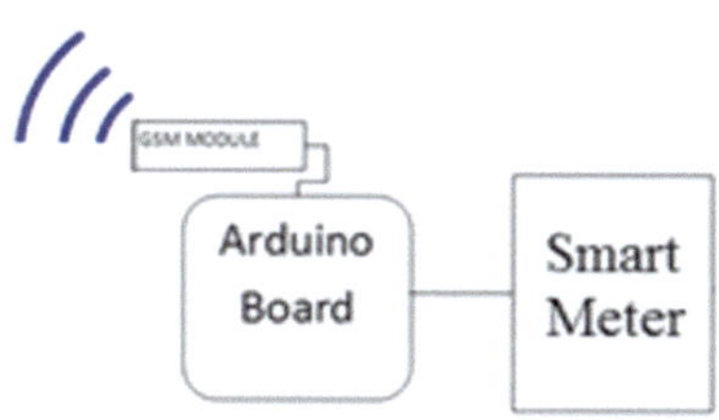

3.2 Smart Line Meters (SLM)

Refer Fig. 3 for the basic structure of a Smart Line Meter.

3.3 Arduino

Arduino is a open source hardware and software architecture that is a microcontroller, it can control the smart devices connected to it such as sensors, GSM, Bluetooth, etc., this hardware is widely used all over the physical world and implements the hardware in an efficient way, it is of ATMEL 8-bit AVR microcontroller that is facilitated with complete components connected to its board [12]. Arduino is a preprogrammed device with boot loader that lets the user to upload the programs into the chip flash memory Fig. 4.

The Arduino board is connected to the digital smart meter and readings are sent to GSM module, this GSM signals the power distributing grid and processes the received signal based on the customer id and meter signal. These components work together as one device and manage the power grid and identify the power consumption at time intervals.

3.4 Power Distributing Grid

The power distributing grid which receives the signal from the SHHM and SLM stores the power values of the customers such that each power distributing grid support is providing the service to customer and the data from the smart household meters (SHHM) and smart line meters (SLM) are indexed in the server to process.

Here, the comparison is done between the received data of SHHM and SLM in which the line not equalized will be identified and forwarded to the admin where the line will be easily identified and theft can be prevented.

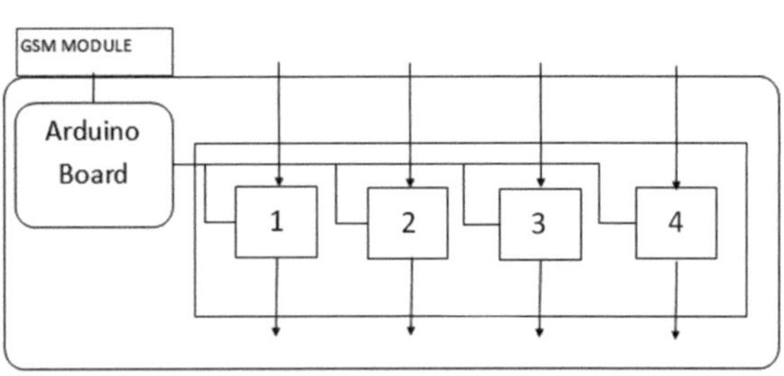

Fig. 3 Smart line meter

Fig. 4 Arduino board

The traveling of electricity loss is 2–5% of overall electricity and it is decreased in the comparison table. The loss indication of more than the 5% percentage will be denoted as theft status.

4 Results and Discussion

We have stimulated the GSM module in omnet++ by slotted aloha overload method and slotted aloha optimal method. While sending the packets to the server which is the local power distributing grid, the processing and sending time is calculated and derived in Fig. 5.

Fig. 5 Slotted aloha overload versus optimal

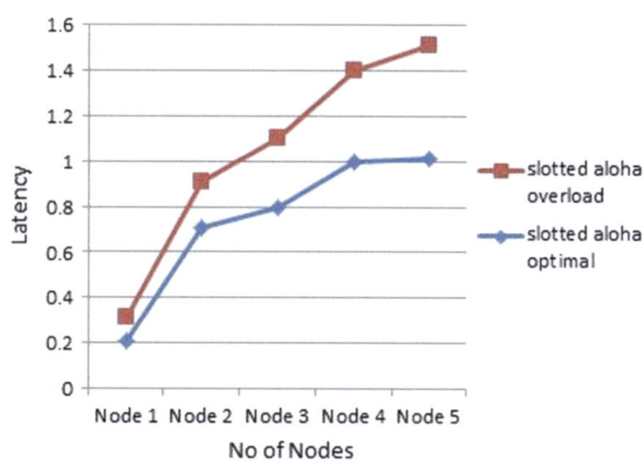

With the proposed model, power theft identification can be found imminently and can be stopped before the high loss of power. Here, the architecture gives the smart grid a flexible and scalable monitoring system and management system where all the processes are automated through the server and this reduces human power which is a benefit to the government.

5 Conclusion

The smart grid technologies are developing in many countries by establishing the infrastructure in their smart cities; smart grid power management system provides the information of all the connections in the networks and with technology, smart devices, sensors, and smart meter. Thus, managing the power grid in an efficient way and identifying the power theft imminently are the most important features of the architecture and the smart grid becomes more secure and managing the system will be flexible and scalable.

References

1. Khan E, Adebisi B, Honary B (2013) Location based security for smart grid applications. The mediterranean green energy forum 2013, MGEF-13. Energy Procedia 42:299–307
2. Rice EB, AlMajali A (2014) Mitigating the risk of cyber attack on smart grid systems. In: Conference on systems engineering research (CSER 2014). Procedia Compute Sci 28: 575–582
3. Bekara C (2014) Security issues and challenges for the IoT-based smart grid. International workshop on communicating objects and machine to machine for mission critical applications (COMMCA-2104)
4. Ashok A, Hahn A, Govindarasu M (2014) Cyber physical security of Wide-Area monitoring, protection and control in a smart grid environment. Dept Electr Comput Eng Iowa State Univ Ames IA USA J Adv Res 5:481–489
5. Yan Y, Qian Y, Sharif H, Tipper D (2012) A survey on cyber security for smart grid communications. IEEE Commun Surv Tutorials 14(4), Fourth quarter
6. Baig ZA, Amoudi A-R (Aug 2013) An analysis of smart grid attacks and countermeasures. J Commun 8(8)
7. Kalaivani R, Gowthami M, Savitha S, Karthick N, Mohanvel S (Feb 2014) GSM based electricity theft identification in distribution systems. Int J Eng Trends Technol (IJETT) 8(10)
8. Federal Energy Regulatory Commission (2013) Assessment of demand response and advanced metering staff report
9. Anderson R, Fuloria S (2010) Who controls the off switch? In: IEEE international conference on smat grid communcations https://doi.org/10.1109/SMARTGRID.2010.5622026
10. Woody C (2013) Mission thread security analysis: a tool for systems engineers to characterize operational security behavior. INCOSE Insight 16(2):37–40

11. Liu Y, Ning P, Reiter MK (2009) False data injection attacks against state estimation in electric power grids. In: Proceedings of the 16th ACM conference on computer and communications security, CCS'09, ACM, New York, USA
12. Sridhar S, Manimaran G (2010) Data integrity attacks and their impacts on SCADA control system. In: Proceedings of power and energy society general meeting

Voltage Stability Analysis for Planning and Operation of Power System

Akhilesh A. Nimje, Pankaj R. Sawarkar and Praful P. Kumbhare

Abstract Voltage control is an important phenomenon in the ever-escalating power system. It is presumed that the voltages at various buses are within their tolerable limits. The elements of transmission line absorb the reactive power. The loads are mostly inductive which influence the voltage profile at buses. The paper examines the voltage profile at various loading conditions and suggests the ways to improve it. The paper studies the requirement of reactive power in a power system.

Keywords Stability · Load flow · Disturbance · Compensation FACTS

1 Introduction

With the expansion of power system in terms of capacity addition and increasing load demand, the researchers have shifted their focus more on voltage stability issues. In the developed countries, where smart grid has been in operation, the concept of power quality is given the utmost importance. The energy charges to be paid by the consumers are the measure of voltage stability and power quality. This has posed several challenges for the utilities to keep the power quality as per the predefined standards before connecting it onto the point of common coupling. Poor voltage regulation in suburban and rural areas of consumers is the consequence of growing power demand and limited capacity addition. Most of the costly equipments either in residential apartments and industrial areas are required to be switched off in the event of poor voltage regulation and voltage flicker. Such problems

A. A. Nimje (✉) · P. R. Sawarkar · P. P. Kumbhare
Electrical Engineering, Guru Nanak Institutions, Nagpur, India
e-mail: nimjeakhilesh29@gmail.com

P. R. Sawarkar
e-mail: pankaj.sawarkar@gmail.com

P. P. Kumbhare
e-mail: praf369@gmail.com

A. Mishra et al. (eds.), *Silicon Photonics & High Performance Computing*,
Advances in Intelligent Systems and Computing 718,
https://doi.org/10.1007/978-981-10-7656-5_3

have scaled up recently due to the inclusion of nonlinear loads. The phenomenon such as voltage stability and voltage collapse needs a closed monitoring and corrective actions before it escalates. The proper planning and operation of power system is thus the sole responsibility of power system analysts. This has initiated the interests of a large number of scientists and engineers to solve the voltage stability problem and improve the power quality by restructuring the generation, transmission, and distribution of electric power systems. The voltage stability problem has been identified by the dynamic behavior of the power system. On the other hand, the load changes are the reasons behind the voltage collapse. The other driving force is reactive power generation in alternators. The computational procedures [1] (Load flow) find applications during the initial planning stages of voltage stability endangered power system or during the development phases of various countermeasures. The maintenance of adequate system voltage profiles in the event of small, medium, and large disturbances has been a matter of major concern [2]. Load flow analysis and transient stability analysis are used as a tool to determine the voltage security.

2 Power System Model

A power system model is generally nonlinear and is described by algebraic and differential equations [3].

$$\dot{X} = F(X, Y, P) \tag{1}$$

$$0 = G(X, Y, P), \tag{2}$$

where $X = (\delta,\ \omega,\ E_d^{'},\ E_q^{'},\ E_{\text{fd}},\ V_R,\ R_f,\ V$ and θ at load bus); $Y = (I_d,\ I_q,\ V,\ \theta)$; $P = (P_{\text{L}}/Q_{\text{L}})$.

A change in parameters of Eqs. (1) and (2) brings the corresponding change in X and the eigen values. Near an equilibrium point, $\dot{X}$ becomes zero. Thus,

$$0 = F(X, Y, P) \tag{3}$$

Linearising Eqs. (1) and (2) at $(X(P_o), Y(P_o))$ as follows:

$$\begin{bmatrix} \Delta\dot{X} \\ 0 \end{bmatrix} = \begin{bmatrix} \frac{\partial F_i}{\partial X_j} & \frac{\partial F_i}{\partial Y_j} \\ \frac{\partial G_i}{\partial X_j} & \frac{\partial G_i}{\partial Y_j} \end{bmatrix} \begin{bmatrix} \Delta X \\ \Delta Y \end{bmatrix}$$

If det $\left(\frac{\partial G_i}{\partial X_j}\right)$ is not zero, then $\Delta\dot{X} = A\Delta X$, where $A = \left[\frac{\partial F_i}{\partial X_j} - \frac{\partial F_i}{\partial Y_j}\frac{\partial G_i}{\partial Y_j}^{-1}\frac{\partial G_i}{\partial X_j}\right]$

3 Voltage Stability

A static voltage stability analysis refers to the solutions obtained in Gauss–Seidel or Newton–Raphson method of power flow. However, the dynamic voltage stability is assessed by modeling the power system incorporating all mechanisms such as swing, flux, excitation, generator load model, etc. At PV bus, P and V are specified whereas Q and δ are unspecified. Here, Q and δ are updated using GS method. It includes the following steps:

Step 1: $Q_i = -\mathrm{Im}\left\{ V_i^* \sum_{k=1}^{n} Y_{ik} V_k \right\}$

Step 2: The revised value of δ is obtained from Step 1. Thus, $\delta_i^{(r+1)} = V_i^{(r+1)}$

$$= \text{Angle of} \left[\frac{A_i^{(r+1)}}{\left(V_i^{(r)}\right)^*} - \sum_{k=1}^{i-1} B_{ik} V_k^{(r+1)} - \sum_{k=i+1}^{i-1} B_{ik} V_k^{(r)} \right].$$

This gives a range of reactive generation, i.e., $Q_{\min}$ to $Q_{\max}$, If Q_i is not within this limit that particular i^{th} bus is then treated as "PQ" bus. In dynamic voltage stability analysis, the machine currents prior to disturbance are calculated [4]. $E'_{i(0)} = E_{ti} + r_{ai} I_{ti} + j x'_{di} I_{ti}$, where $E'_{i(0)} = e'_{i(0)} + j f'_{i(0)}$ and $\delta_{i(0)} = \tan^{-1} \frac{f'_{i(0)}}{e'_{i(0)}}$; $e'^{(1)}_{i(t+\Delta t)} = |E'_i| \cos \delta^{(1)}_{i(t+\Delta t)}$ and $f'^{(1)}_{i(t+\Delta t)} = |E'_i| \sin \delta^{(1)}_{i(t+\Delta t)}$.

4 Numerical Investigation

A preliminary numerical investigation was performed on π model of a radial line of length 250 km, 400 kV, three phase.

Test Data [5]

Line length = 250; Frequency in Hz = 50; r = 0.014 Ω/ph/km; L = 0.95 mH/Ph/km; C = 0.01 μF/Ph/km; Conductance g = 0 S/ph/km. Receiving end (L-L) voltage kV = 400 ∠0°; P_r = 600 MW; Q_r = 450 MVAR.

4.1 Equivalent π Model

Z' = 3.43193 + j 73.8878 ohms; Y' = 1.82042e−007 + j 0.000789256 mho; Zc = 308.305 + j−7.22715 Ω; alpha 1 = 0.00567619 neper; beta 1 = 0.242143 rad = 13.8737°; A = 0.97084 + j 0.0013611; B = 3.4319 + j 73.888; C = −3.5773e−007 + j 0.00077775; D = 0.97084 + j 0.0013611.

4.2 Line Performance for a Given Load

Vr = 400 kV ∠0° (Line); Pr = 600 MW; Qr = 450 Mvar; Ir = 1082.53 A ∠−36.8699°; cos ϕ_r = 0.8 lag; Vs = 488.585 kV (Line) ∠12.7122°; Is = 954.232 A ∠−28.1227°; cos ϕ_s = 0.756597 lag; Ps = 610.969 MW; Qs = 528.024 Mvar; PL = 10.969 MW; QL = 78.024 Mvar; %VR = 25.8146%; η = 98.2047%.

4.3 Line Performance for a Given Source

Vs = 420 kV ∠0° (Line); Ps = 450 MW; Qs = 300 Mvar; Is = 743.452 A ∠0° −33.6901°; cos ϕ_s = 0.83205 lag; Vr = 359.457 kV ∠−12.229° (Line); Ir = 841.028 A ∠−44.3699°; cos ϕ_r = 0.846746 lag; Pr = 443.375 MW; Qr = 278.565 Mvar; PL = 6.625 MW; QL = 21.435 Mvar; %VR = 20.3521%; η = 98.5278%.

4.4 Line Performance for Specified Load Impedance of 250 + j * 0 Ohms Per Phase

Vr = 400 kV ∠0°(Line); Ir = 923.76 A ∠0°; cos ϕ_r = 1; Pr = 640 MW; Qr = 0 Mvar; Vs = 411.346 kV ∠16.781°(Line); Is = 914.802 A ∠11.4034°; cos ϕ_s = 0.995598 lag; Ps = 648.902 MW; Qs = 61.089 Mvar; PL = 8.902 MW QL = 61.089 Mvar; %VR = 5.92495%; η = 98.6282%.

4.5 Line Performance at no Load and Determination of Shunt Compensation

Vs = 400 kV ∠0°(Line); Vr = 412.013 kV ∠−0.00140194°(Line); Is = 185.008 A ∠89.946°; cos ϕ_s = 0.00094199 lead; V_R = 400 kV; Shunt reactance = 2534.03 Ω; rating = 63.1405 Mvar.

4.6 Line SC at Load

Vs = 400 kV ∠0°(Line); Ir = 3122.18 A ∠−87.3406°; Is = 3031.15 A ∠−87.2603°.

4.7 For Shunt Capacitive Compensation

V_S = 400 kV; V_R = 400 ∠0° kV; P_r = 600 MW; Q_R = 450 MVAR.

Vs = 400 kV ∠16.2325° (Line); *Vr* = 400 kV ∠0° (Line); P_{load} = 600 MW; Q_{load} = 450 Mvar; Load current = 1082.53 A ∠−36.8699°; cos ϕ_l = 0.8 lag; Required shunt capacitor: 319.327 Ω, 8.30679 μF, 501.054 Mvar; Shunt capacitor current = 723.209 A ∠90°; *Pr* = 600.000 MW; *Qr* = −51.054 Mvar; *Ir* = 869.155 A ∠4.86354°; ϕ_r = 0.996399 lead; *Is* = 877.648 A ∠16.709°; cos ϕ_s = 0.999965 lead; *Ps* = 608.031 MW; *Qs* = −5.057 Mvar; PL = 8.031 MW; QL = 45.997 Mvar; %VR = 3.00326%; η = 98.6792%.

4.8 Series Compensator

Vr = 400 kV ∠0° (Line); *Pr* = 600 MW; *Qr* = 450 Mvar; %comp = 50%; Required series capacitor: 36.9439 Ω, 71.8003 μF, 39.2281 Mvar; Subsynchronous resonant frequency = 35.3553 Hz; *Ir* = 1082.53 A ∠−36.8699°; PF*r* = 0.8 lag; *Vs* = 443.946 kV ∠6.73932° (Line); *Is* = 969.205 A at −28.1955° cos ϕ = 0.819804 lag; *Ps* = 610.965 MW; *Qs* = 426.767 Mvar; PL = 10.965 MW; QL = −23.233 Mvar; %VR = 12.6283%; η = 98.2053%.

4.9 Compensation Requirements

V_S = 400 kV; V_R = 400 ∠0° kV; *P*load = 600 MW; *Q*load = 450 Mvar; Load current = 1082.53 A ∠−36.8699°; cos ϕ = 0.8 lag; Shunt capacitor: 329.45 Ω, 8.05154 μF, 485.658 Mvar; Capacitor current = 700.986 A ∠90°; Series capacitor: 36.9439 Ω, 71.8003 μF, 28.4606 Mvar; Subsynchronous resonant frequency = 35.3553 Hz; *Pr* = 600 MW; *Qr* = −35.6575 Mvar; *Ir* = 867.553 A ∠3.40104°; cos ϕ_r = 0.998239 lead; *Is* = 884.446 A ∠152633°; cos ϕ_s = 0.992165 lead; *Ps* = 607.961 MW; *Qs* = −76.557 Mvar; PL = 7.961 MW; QL = −40.899 Mvar; %VR = 1.47935%; η = 98.6906% (Figs. 1, 2, 3 and 4).

5 Discussion and Conclusion

For a given receiving end quantities at lagging power factor loads, P_S and Q_S are always higher than that of the receiving ends. This implies that the additional required reactive power is supplied by the generators. There are active power losses in the transmission section which can be minimized with the proper sizing of conductors. The angular difference between V_S and V_R depends upon the

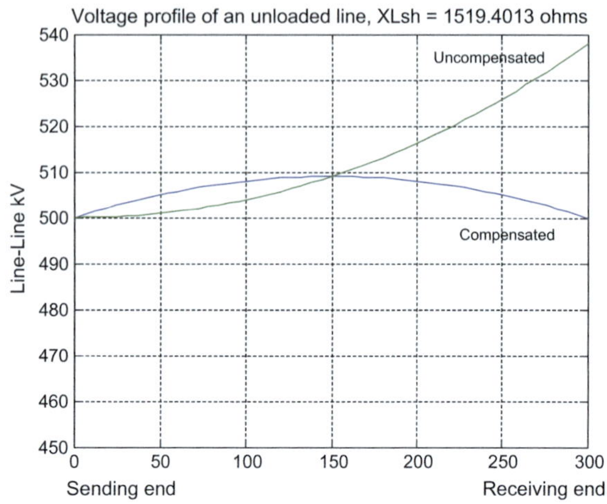

Fig. 1 Voltage profile of an unloaded line

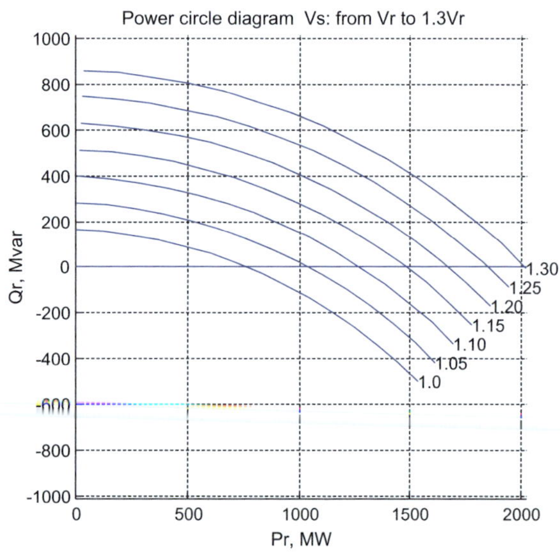

Fig. 2 Receiving end power circle diagram

transmission line reactance and the load power factor. If the voltage profile worsens due to some of the reasons, the determination of shunt compensation helps to restore and maintain the voltage within its tolerable limits. For an unloaded line, the receiving end voltage is found more as compared with the sending end voltage due to line charging capacitance. Or in other words, it can be said that on no load, the capacitive reactance dominates the inductive reactance. If the line is short circuited at the receiving end, the current reaches abruptly a very high value and lags behind the voltage approximately by 90°. A series capacitor compensation may be employed to bring down the transfer reactance between the buses. This helps in

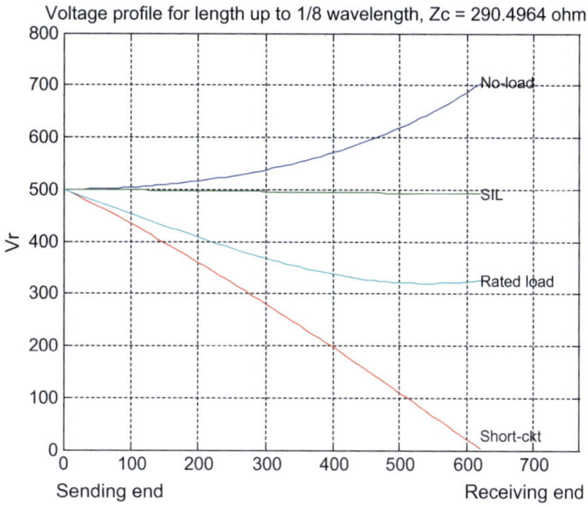

Fig. 3 Voltage profile

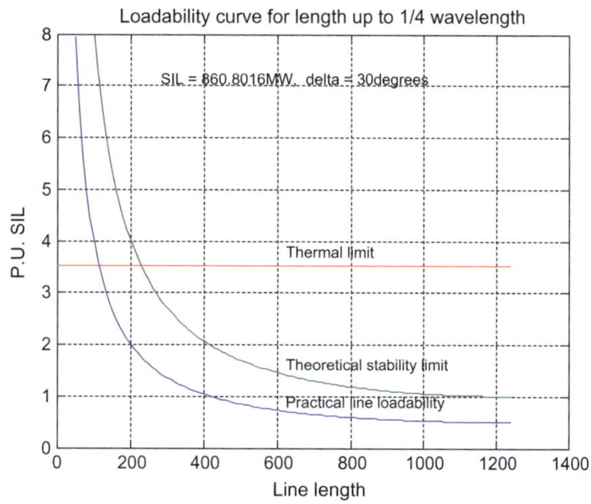

Fig. 4 Loadability curve

enhancing the static transmission capacity. The recommended percentage of series compensation is limited to 25–30% as it would be uneconomical to go beyond 30% due to economical constraints. The compensation may be provided with various configurations (series and parallel) of capacitors and inductors for the purpose of voltage control and maintain the power quality. There has been a tremendous advancement in this field with the Flexible AC Transmission System (FACTS) controllers. The voltage stability and power quality problems have been greatly resolved with FACTS controllers.

6 Future Work

Based on the preliminary numerical investigations as presented in Sect. 4 of this paper, the work can be extended to fabricate a scale model of IEEE 9 bus system in institute laboratory. The transmission model of π configurations would be developed for a balanced three-phase system [6]. With loads connected at bus 5, 6, and 8 and generations at 1, 2, and 3, the experimental and numerical results obtained with Gauss–Seidel Method [7, 8, 9] will be compared for small variations in loads at various buses. The voltages at various buses are required to be within the tolerable limits [10, 11, 12]. If it violates, the appropriate compensations with RC loads would be applied to compensate for voltages (Figs. 5 and 6).

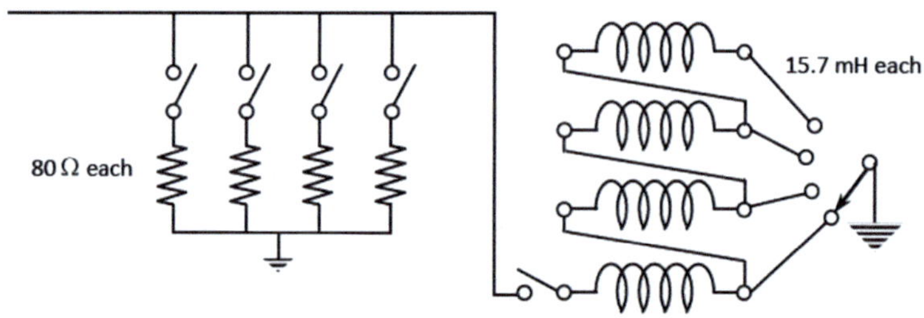

Fig. 5 R-L load

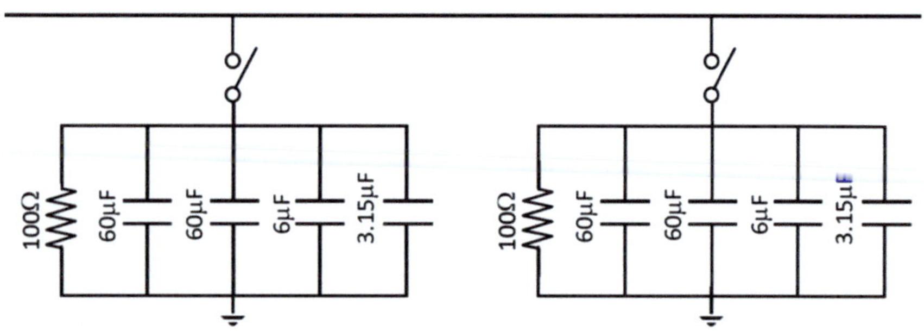

Fig. 6 R-C load

Acknowledgements The authors are grateful to the management of the Guru Nanak Institute of Engineering & Technology, Nagpur for providing necessary support and infrastructure to work.

References

1. Nagsarkar TK, Sukhija MS (2007) Power system analysis. First Edition, Oxford University Press, New Delhi
2. Van Cutsem T, Hacquemart Y, Marquet JN, Pruvot P (Aug 1995) A comprehensive analysis of mid term voltage stability. IEEE Trans Power Syst 10(3)
3. Lee B, Ajjarapu V (Nov 1995) A piecewise global small disturbance voltage stability analysis of structure preserving power system models. IEEE Trans Power Syst 10(4)
4. Nimje AA, Panigrahi CK (Dec 27–29, 2009) Transient stability analysis using modified Euler's iterative technique. Third international conference on power systems (ICPS), Kharagpur, India
5. Saadat H (2006) Power system analysis. Tata McGraw-Hill Publishing Company Limited, New Delhi, Eighth Edition
6. Terzioglu R, Cavus TF (Sept 2013) Probablistic load flow analysis of the 9 bus WSCC system. Int J Sci Res Publ 3(9):ISSN 2250–3153
7. Mienski R, Pawelek R, Wasiak I (March 2004) Shunt compensation for power quality improvement using a STATCOM controller: modelling and simulation. IEE Proc Gener Transm Distrib 151(2):274–280
8. Tamrakar I, Shilpakar LB, Fernandes BG, Nilsen R (2007) Voltage and frequency control of parallel operated synchronous generator and induction generator with STATCOM in micro hydro scheme. IET Gener Transm Distrib 1(5):743–750
9. Padiyar KR (2007) FACTS controller in power transmission and distribution. First Edition, New Age International Publishers, New Delhi
10. Hingorani NG, Gyugyi L (2001) Understanding FACTS. IEEE power engineering society, Sponsor, IEEE Press, Standard Publishers Distributors, Delhi
11. Overbye TJ (May 1993) Use of energy methods for online assessment of power system voltage security. IEEE Trans Power Syst 8(2)
12. Kirby B, Hirst E (Dec 1997) Ancillary service details: voltage control, an approach to corrective control of voltage instability using simulation and sensitivity. Oak, Ridge National Laboratory (Energy Division) sponsored by National Regulatory Research Institute Columbus, Ohio

Application of Distributed Static Series Compensator for Improvement of Power System Stability

Praful P. Kumbhare, Akhilesh A. Nimje and Pankaj R. Sawarkar

Abstract This paper is based on the basic performance analysis of Distributed Static Series Compensator (DSSC) required for the improvement of power system stability. The DSSC has small size and weight with cylindrical structure that is put on the line conductor without grounding. It has the capability of changing the impedance and altering the power flow through the lines. The fundamental behavior of DSSC and Static Synchronous Series Compensator (SSSC) is similar. However, a DSSC costs less and higher reliability. In this paper, single machine infinite bus system (SMIB) has been simulated in PSCAD for various operating conditions of power system such as steady state and during fault conditions.

Keywords D-FACTS · DSSC · FACTS · PSCAD · SMIB · Stability

1 Introduction

Modern power system has an interconnected grid network that covers generation, transmission, and distribution. The generating stations are located far away from the consumers with a transmission network that requires high degree of reliability to facilitate the uninterrupted power supply at load centers. Expansion of the transmission system requires huge investment and hence the power system engineers attempt to utilize the full capacity of the existing transmission lines. The term Flexible Alternating Current Transmission System (FACTS), in simple words, means applying flexibility to electric power system. It refers to the ability to accommodate changes in the electric transmission system or operating conditions

P. P. Kumbhare (✉) · A. A. Nimje · P. R. Sawarkar
Electrical Engineering, Guru Nanak Institutions, Nagpur, India
e-mail: praf369@gmail.com

A. A. Nimje
e-mail: nimjeakhilesh29@gmail.com

P. R. Sawarkar
e-mail: pankaj.sawarkar@gmail.com

A. Mishra et al. (eds.), *Silicon Photonics & High Performance Computing*,
Advances in Intelligent Systems and Computing 718,
https://doi.org/10.1007/978-981-10-7656-5_4

while maintaining sufficient steady state and transient margins. The power system stability plays a vital role in power system as the load keeps on changing. A step change in the input voltage may cause small signal disturbance while the three phase faults lead to large disturbance in power system. Thus, stability is defined as the ability of power system to regain synchronism after being subjected to some form of disturbance within least possible time [1]. FACTS technology is the best option for today's power systems scenario for the improvement of power flow in transmission lines, maintaining voltage profile and improving system capability by damping the power oscillations. FACTS technology has a family of controllers with various configurations such as series, shunt, series–series, and series–shunt.

However, widespread use of this technology is restricted due to increase in cost and poor reliability. The device requirements of more component and complexity increase the cost of installation while the single-point failure may shut down the whole system. These limitations are overcome by of the use of Distributed FACTS (D-FACTS) devices [2].

2 Distributed FACTS

Distributed Flexible A.C Transmission system is an advanced version of FACTS system which not only improves the power transmission capability but also improves the reliability and security within the permissible cost. It reduces the power flow through overloaded lines, minimizing losses and cost. The weak and low stability networks required number of modules of D-FACTS devices. The various D-FACTS controllers are distributed series impedance, distributed series reactor, and distributed static series compensator.

2.1 Distributed Static Series Compensator

A DSSC is a voltage-sourced converter and has the objective to provide compensation to the line by improving the voltage across the impedance of respective transmission line. This helps to increase the power flow and current. It consists of a single-turn transformer (STT), single-phase inverter, and a controller. STT has higher turns ratio as a result of which the current through the inverter is reduced and IGBTs can easily used here for reducing the cost [3]. The primary winding of STT is connected with the inverter and the secondary is the transmission line. When DSSC is clamped on the transmission line, a core magnetic circuit is formed around the primary and secondary winding of transformer. Each module consists of a control circuit with communication system in order to coordinate the operation. When DSSC starts, the inverter voltage and line current are in phase, only the real power is extracted from the transmission line to charge the DC bus capacitor. It injects the reactive impedance or quadrature voltage in series with the transmission line. Thus, there is increase or decrease in power along the line [4]. The feasibility of DSSC has been tested in PSCAD and the power oscillations obtained during fault have been damped satisfactorily.

2.2 *Role of DSSC for Power Flow Improvement*

Figure 1 shows two parallel lines (20 and 30 miles) having voltage level 132 kV. The impedance is (0.17 + *j*0.8) ohm/miles for each line Tables 1 and 2.

From the above calculation, it has been concluded that about 20% power flow is improved by providing compensation (or connecting DSSC) to the transmission line [5].

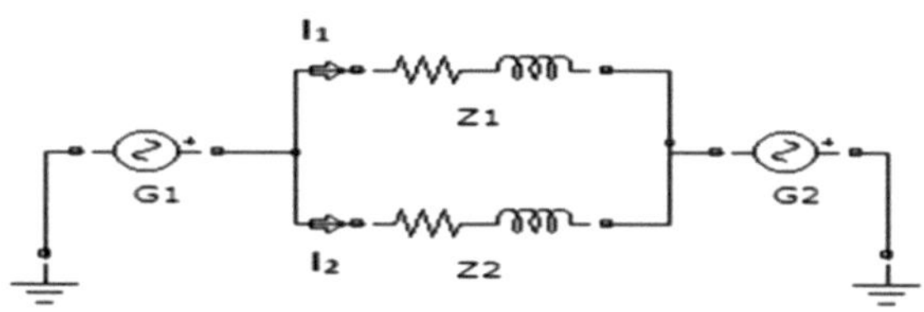

Fig. 1 Two-generator parallel line circuit

Table 1 Given data

(a) Test data	
Without DSSC	With DSSC (A variation of ± 20% in line impedance)
G1 = 132° kV, $\delta 1 = 7.95°$	G1 = 132.0° kV, $\delta 2 = 9.48°$
G2 = 1327.95° kV	G2 = 1329.48° kV
Z1 = (3.4 + *j*16) Ω = 16.35778°	Z1 = [(3.4 + *j*16) + *j*3.2] Ω = 19.580°
Z2 = (5.1 + *j*24) Ω = 24.53578°	Z2 = [(5.1 + *j*24) − *j*4.8] Ω = 198,675.124°

Table 2 Improvement of power and current by injecting impedance

(b) Calculation	
(i) Without DSSC	(ii) With DSSC
$P_1 = \frac{V_1 V_2}{Z_1} \sin\delta = \frac{132^2}{16.35} \sin(7.95) = 147\,\text{MW}$	$P_1 = \frac{V_1 V_2}{Z_1} \sin\delta = \frac{132^2}{19.5} \sin(9.48) = 147\,\text{MW}$
$I_1 = \frac{2v\sin\left(\frac{\delta}{2}\right)_1}{Z_1} = \frac{2(132)\sin\left(\frac{7.95}{2}\right)}{16.35} = 644\,A$	$I_1 = \frac{2v\sin\left(\frac{\delta}{2}\right)_2}{Z_1} = \frac{2(132)\sin\left(\frac{9.48}{2}\right)}{19.5} = 644\,A$
$P_2 = \frac{V_1 V_2}{Z_2} \sin\delta = \frac{132^2}{24.54} \sin(7.95) = 98\,\text{MW}$	$P_2 = \frac{V_1 V_2}{Z_2} \sin\delta = \frac{132^2}{19.86} \sin(9.48) = 145\,\text{MW}$
$I_2 = \frac{2v\sin\left(\frac{\delta}{2}\right)_1}{Z_2} = \frac{2(132)\sin\left(\frac{7.95}{2}\right)}{24.54} = 430\,A$	$I_2 = \frac{2v\sin\left(\frac{\delta}{2}\right)_2}{Z_2} = \frac{2(132)\sin\left(\frac{9.48}{2}\right)}{19.86} = 632\,A$

iii) Power flow between two buses

	V (kV)	XI (Ohm)	Ohm	Ohm	Degree	I (amp)	*P (MW)*
Line1 initial	132	16	0	16	7.95	644	*147*
Line2 initial	132	24	0	24	7.95	430	*98*
Line1 + DSSC	132	16	+3.2	19.2	9.48	644	*147*
Line2 + DSSC	132	24	−4.8	19.2	9.48	632	*145*

3 System and Controller Details

3.1 System Representation

Figure 2 shows a generator connected to infinite bus. Each DSSC injects maximum 2 V into the line through single-turn transformer of 1:75 ratios.

Figure 3 shows the simulated DSSC module. Each transmission line is connected with three DSSC modules which increase the active power flowing through a transmission line by 0.6 MW [6, 7]. Again, a three-phase line to ground fault is activated at generator side for measuring the impact of DSSC for damping the oscillation produced due to fault.

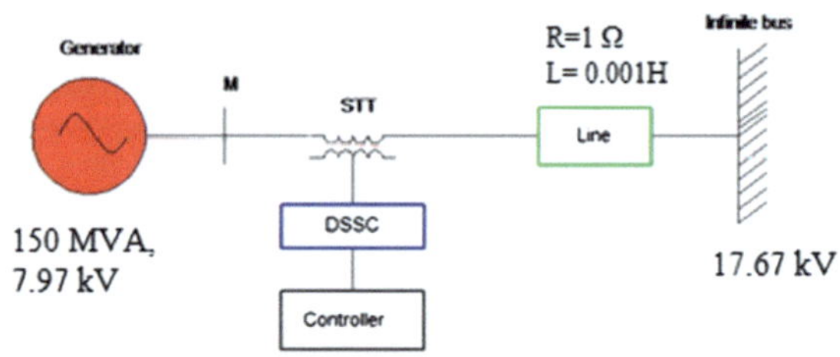

Fig. 2 SMIB with DSSC

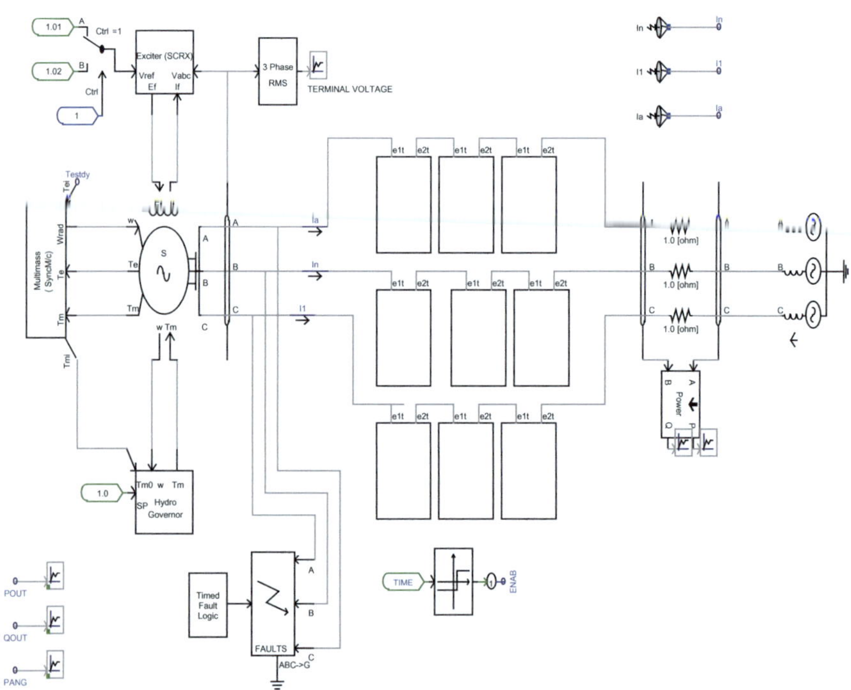

Fig. 3 System model in PSCAD with DSSC

3.2 Control Model

The primary function of DSSC is to control the flow of power in transmission line. This requirement can be achieved by providing control to the system directly or indirectly [8, 9, 10]. There are few limitations of direct control such as complexity in circuit, increased losses, and increased harmonic content. Hence, it is recommended to use indirect controller [11, 12]. The error signal is generated by comparing dc voltage with reference voltage and is given to PI controller and it generates the necessary phase displacement. The phase-locked loop (PLL) gives the basic synchronization angle θ which is the phase angle of system line current. The obtained angle from PI and PLL is given to PWM inverter so control firing of IGBTs is obtained. A filter is provided in the control circuit for reducing the ripples from the system (Figs. 4 and 5).

Controller shown in Fig. 5 is designed in PSCAD. A signal g1p, g1n, g2p, and g2n generates from controller and are used to generate the firing signals for converter. With the variation in firing angle, the injected voltage is controlled.

4 Simulation and Result

The PSCAD model of single generator infinite bus system shown in Fig. 3 is simulated in the steady state with and without DSSC. The transmitted power flow in line is increased by connecting a number of DSSCs.

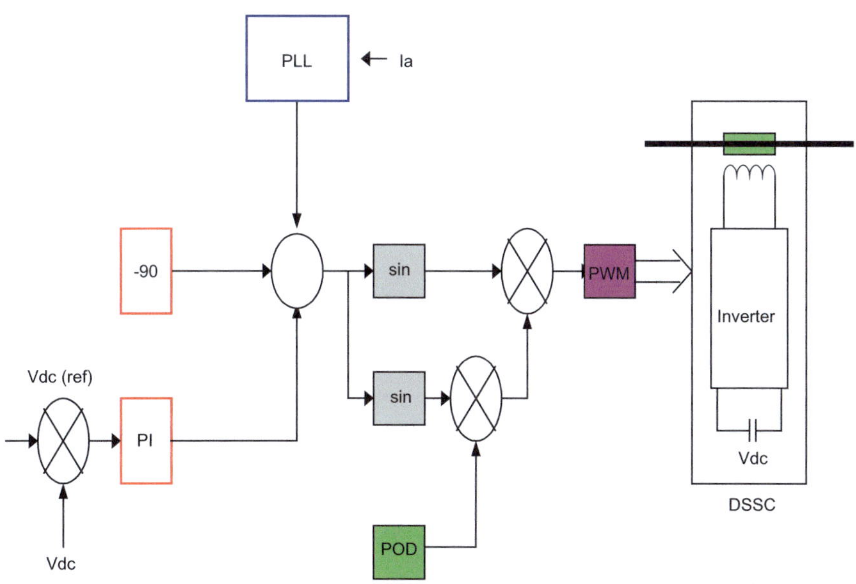

Fig. 4 Control model of DSSC

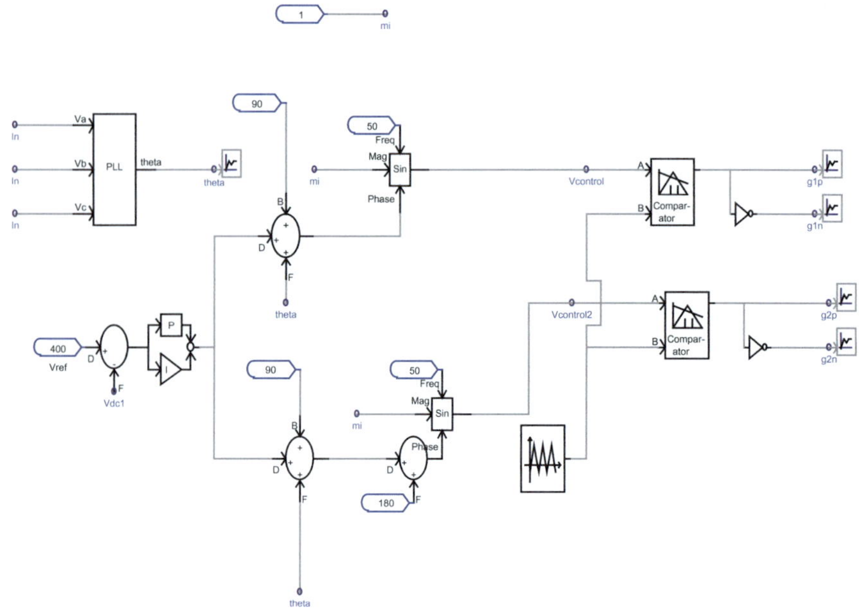

Fig. 5 Control scheme of DSSC in PSCAD

4.1 Steady-State Operating Condition

The system shown in Fig. 2 is a single-generator infinite bus designed in PSCAD software which is simulated considering the same parameter of Sect. 3. Active power measures 85 MW at steady state (without fault) when DSSC is out of system. After connecting a single DSSC in each line of a system, the power increased up to 0.2 MW. In a complete system, 9 number of DSSC are connected, that generates 1.8 MW of power (Fig. 6).

4.2 Dynamic Condition (Considering Power Angle of Generator)

To understand the dynamic performance of DSSC, a symmetrical fault is created at 1.0–1.05 s on the generator side. Figure 7 shows the power oscillation with and without DSSC.

Without DSSC	With DSSC
Power angle escalates up to 124.07° at 1.010 s, when fault is done on 1.00 s. After fault clearance, it is settled at 33°	Power angle escalates to 108.07° at 1.010 s, when fault is done on 1.00 s. After fault clearance, it settles at 33°

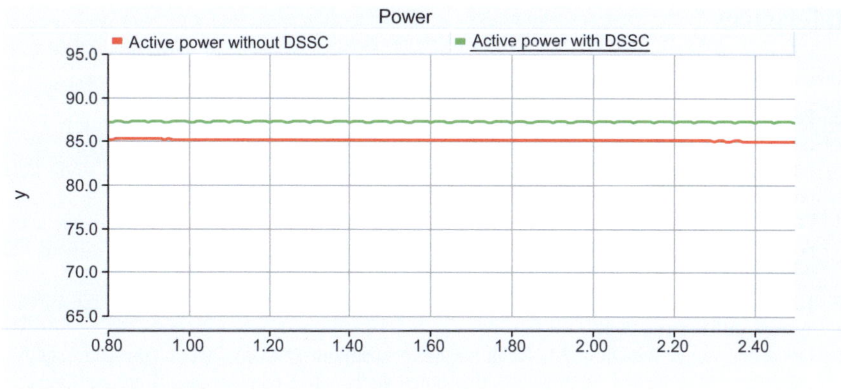

Fig. 6 Active power flows in line with and without DSSC

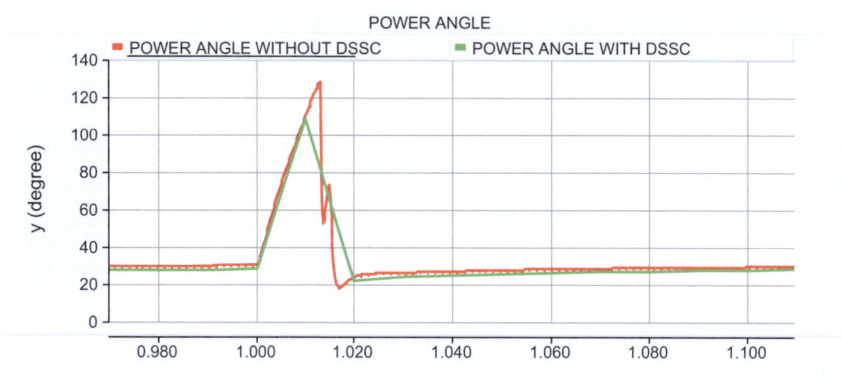

Fig. 7 Comparison of power angle of generator with DSSC and without DSSC

The above simulation has 09 number of DSSC connected in line which damped up to 16 degree. Hence, it is seen that the power oscillation damps faster with DSSC as compared to without DSSC.

5 Conclusion and Future

In order to penetrate large chunk of transmission and distribution system market, it is necessary to reduce cost and improve reliability of existing FACTS devices. DSSC is a member of D-FACTS family that improves power transfer capability and system stability. As of now, DSSCs are under development and few prototypes have been reported. More research is required to justify its commercial applicability.

References

1. Kundur P (1994) Power system stability and control. Prentice Hall NY, U.S.A
2. Divan D, Gohal H (Nov 2007) Distributed FACTS—a new concept for realizing grid power flow control. IEEE Trans 22:2253–2260
3. Fajri P, Afshamia S (2008) Modeling simulation and group control of distributed static series compensator. American engineering and applied science pp 347–357
4. Divan DM (Oct 2004) A distributed static series compensator system for realizing active power flow control on existing power flow control. Power Syst Conf Exposition, IEEE PES 2:654–661
5. Hingorani NG Understanding FACTS: concepts and technology of flexible AC transmission system. IEEE press 2000
6. Golshannavaz S, Mokhtari M, Khalilian M, Naazarpour D (May 2011) Transisent stability enhancement in power system with distributed static series compensator. Irani conference
7. Simfukwe D, Pal BC, Begovic M, Divan D (July 26–30 2009) Control of power system static stability using distributed static series compensator. Power and energy society general meeting, Calgary, AB, Canada
8. Divan DM Control of power system static stability using distributed static series compensator. IEEE power & energy society general meeting, PES '09
9. Gygui L, schaudar C Sen K (Jan 1997) Static synchronous series compensator: a solid state approach to series compensation of transmission line. IEEE Trans power Delivery 12 n1:406–407
10. Divan D, Johal H (Nov/Dec2007) Design Consideration of series connected distributed FACTS converters. Ind Appl Conf 43(6)
11. Afsharnia S (03/2008) A PSCAD/EMTDC model for distributed static series compensator (DSSC). 2008 second international conference on electrical engineering
12. Abhishek R, Sadda A, Nebhanani L, Maheshwari VM, Pareek VS Optimul allocation of distributed static series compensator in power system using biogeoraphy based optimization. 2012 IEEE 5th India

A Novel Transmission Power Efficient Routing in Cognitive Radio Networks Using Game Theory

Sonia Garg, Poonam Mittal and Chander Kumar Nagpal

Abstract Owing to frequently increasing demand of wireless communication technology, the problem of spectrum shortage arises. To overcome this, Cognitive Radio (CR) came into play. CRs use the available vacant spectrum of primary users intelligently. CRs use these spectrum holes opportunistically by changing their transmission parameters. To model the performance of a wireless network, game theory has been used due to its capability to model individual, independent decision-makers. Game theory can be used in any network at various layers, to model its behavior and performance. To send data between any two nodes in network, we need a routing protocol. We aim to find out a transmission power-aware routing algorithm which routes the message packets efficiently within the network, while maximizing the overall throughput of the system. And then we compare its performance with shortest path and minimum transmission power routing scheme. Implementation is done in MATLAB-9.0.

Keywords *Cognitive* radio · Wireless network · Game theory
Physical layer · Power allocation Cooperative and Non-cooperative game

1 Introduction

Shortage of spectrum is due to the exponential growth of wireless devices and rigid allocation policies of the spectrum. Whereas large portions of the designated frequency bands are only partially occupied, this leads to inefficient spectrum utilization. So a new technique Cognitive Radio (CR) was proposed [1] which can use

S. Garg (✉) · P. Mittal · C. K. Nagpal
Computer Engineering, YMCAUST, Faridabad, India
e-mail: gargsonia76@gmail.com

P. Mittal
e-mail: poonamgarg1984@gmail.com

C. K. Nagpal
e-mail: nagpalckumar@rediffmail.com

A. Mishra et al. (eds.), *Silicon Photonics & High Performance Computing*,
Advances in Intelligent Systems and Computing 718,
https://doi.org/10.1007/978-981-10-7656-5_5

the radio resources with high intelligence and more capabilities [2]. Cognitive capability can be explained as the capability of a CRN to sense spectrum [3] and capture temporal and spatial variations.

PUs has their licensed fixed spectrum. Cognitive radio transceiver needs to sense the environment for the presence of PUs [4], before starting communicating with one another. Secondary Users (SU) or cognitive users can only use the spectrum for a secure communication with other users only when there is no PU in the communication in that spectrum [5–8].

In a CRN, if there are different types of user, i.e., they are different on the basis of their behavior, objectives, then users might not be entirely cooperative. In a multihop environment, a node will forward another packet; if there is some cooperation between the nodes, only then packets can be forwarded reliably. There are many techniques to deal with cooperation like reputation-based and price-based system but these techniques are not able to calculate cooperation incentives provided by these schemes. To overcome this, techniques are needed which can analyze the user's behavior interactively like game theory.

1.1 Cognitive Radio and Game Theory

Game theory is an advance mathematic tool [9] that helps in analyzing the user's behavior. It analyzes the behavior and makes the decisions accordingly. Game theory can be classified as 1. cooperative game model and 2. non-cooperative game model.

Cooperative games are those games in which CR players cooperate with each other to maximize network utility. In non-cooperative games, players are selfish users and take actions independently aiming to maximize their own utility functions [10]. Game theory aims toward Nash equilibrium, i.e., an optimal combination of strategies of all the present players which are normalized [11]. Main components of a game are described in Table 1.

Table 1 Mapping of cognitive radio network elements to a game

Game component	Comments	Modeled element of CRN
Players	Players aim to maximize their utility function by considering the activity of PUs	Nodes in wireless network
Strategy/Set of actions	Actions are the functionality like in a network action is forwarding a packet	Modulation scheme, coding rate, channel allocation, transmission power level, routing path selection, etc.
Utility function/ Set of preferences [12]	It is the player's objective which is obtained by the behavior of cognitive radios	Performance metrics, e.g., throughput, delay, SINR, QoS, etc.

Game theory models can be used in CRN to deal with resource allocation problem (Channel allocation, power control), trust management, and better understand of several issues. Table 1 shows how different CR elements can be mapped on to a game. In this work, power allocation in game theory is used to collect information related to residual power available at all other nodes.

2 Power Allocation: Literature Review

In CRNs, game theory may be used to model and examine CRN at different layers of OSI model. Thomas et al. [1] shows which CRN games can be applied at different OSI layers. Avoiding interference is the key challenge in CRN at physical layer. It provides good QoS to CRN users [13].

To deal with the communication in multichannel CDMA-based CRN, a non-cooperative game [14] is used for power control. It provides a path in the network where least energy is used for forwarding the data by opportunistically accessing the PU's channels.

To minimize the interference generated by SU, it may affect the spectrum sharing mechanism. Interference may be avoided by using the proposed scheme [15] based on power control. The goal is achieved by using non-cooperative game with pricing-based power control.

To provide the dynamic spectrum sharing among CR users, a non-cooperative game-based power control scheme [16] is used for CDMA-pricing cognitive radio system.

To deal with the decentralization of users, an MC-CDMA cognitive radios system with hand-off technique [17] was proposed for cognitive users. Sigmoid efficiency function and nonlinear pricing are used with non-cooperative game which is related to SINR value of the user. Frequency band of PUs is fixed but it may vary for SU according to the availability of spectrum. So modulation and demodulation do not affect the system.

To deal with the channel allocation and power control in CRN jointly, a game theoretic approach is designed which is based on physical interference [18]. SINR may be considered as a physical interference to establish a link. This is an efficient realistic protocol that can handle opportunistic spectrum access by interaction. This technique is valid only for a small local less scalable system but its performance is comparable with global centralized system.

An iterative method is used for resource allocation if the system has incomplete information. There are two techniques used to deal with this incomplete information system for resource allocation like MQAMI method [19] and game theory based technique. Game theory based technique achieves better results than MQAMI-based results but MQAMI is more distributed than this.

A heuristic technique is used to deal with the spectrum allocation problems [20] which mainly exist when there are two networks Primary User Network (PUN) and Cognitive Radio Network (CRN) which are operating on the same frequency band.

This approach is well suited for distributed network.

3 Proposed Work

In this work, an effort is done to design a routing protocol that aims toward optimizing the QoS parameters of the CR system by considering the constraints like power constraint and interference from and to the licensed band (primary users).

Total power can be defined as the sum of residual energy available at all the nodes. Residual energy is the energy contained in the node, i.e., the energy available at the node for consumption. Transmission energy is the energy required for transmission between two or more nodes. Initially, each node in the network has some fixed residual energy, but as the time passes, communication takes place and some transmission energy is required to send the messages in the network and thus the residual energy decreases. The more be the residual energy of a node, the node will be active for more time in the network. On the other hand, the more distant the node is, the more transmission power will be needed to receive the signal at receiver node. The more transmission power needed implies increased interference to other users in the network. In an obstacle-free path, the transmission power needed to send a signal from node1 to node2 is calculated as

$$P_{tr} \alpha \, (\text{Distance}_{1,2})^2 \tag{1}$$

If we consider the realistic environment having obstacles like rivers, buildings, factories, etc., then fading of signal may be there. In such a scenario, transmission power needed to send a signal from node1 to node2 is calculated as

$$P_{tr} \alpha \, (\text{Distance}_{1,2})^4 \tag{2}$$

Game theory is used to analyze and obtain information about the user behavior, network configuration, and other details. This information is collected through a non-cooperative repeated game theory. One of the ways to collect the information is by hidden and exposed terminal. Another way is that source node first sends a hello packet to all its neighbors and then waits for a reply. And all the neighbor nodes flood this packet throughout the network till it reaches the destination node. In reply to this hello message, all the nodes send an ACKnowledgement (ACK) packet back to the sender of hello message. The ACK contains the details about the residual energy at the node and the position of the node in the network. This information is then utilized by the source node to take decision about the routing path. Based on this information, source finds out all paths available from source to destination node. Then, for every path, calculate the transmission energy required on that path. Then, find out the most cooperative path which requires minimum energy. Game theory model is described in Table 2.

Table 2 Game for reputation system

	Node i		
Node j		Cooperative	Non-cooperative
	Cooperative	(p, c)	U (Ci, Ij)
	Non-cooperative	U (Ii, Cj)	(0, 0)

Here,

$$\begin{aligned} U(C\,i,\ I\,j) = {} & (-c, p) && \text{if}\, R(j) > Tr \\ & (0,\ 0) && \text{if}\ \ R(j) < \ = \ Tr \\ U(I\,i,\ C\,j) = {} & (p,\ -c) && \text{if}\ \ R(i) > Tr \\ & (0,\ 0) && \text{if}\, R\ (i) < \ = Tr \end{aligned}$$

Here, p is the profit gained in forwarding a packet and c is the cost of forwarding a packet. Figure 1 presents an algorithm for proposed routing protocol.

```
1. Fix energy for all nodes.

2. Find SN and DN.   // find source and destination

3. SN broadcast HM.        // to collect behavior of
   neighboring nodes

4. SN receives residual energy, position of node
   from NN. Repeat same process for all NNs until
   HM reaches DN.          // non cooperative game
   is implemented with multiple iterations

5. Find all paths between SN and DN. // multiple
   paths are there with corresponding cooperation
   value

6. Calculate TE for each path.

7. Select path having minimum TE and better
   cooperation

8. If(TE1=TE2)    // two paths require same TE

                 Choose path having maximum RE.

9. If(path Ni,i+!>=RTE) (for all ith nodes on path)
   i.  Route data.
   ii. Calculate available RE at every node; REi=REi-
         RTEi
              Else
   i.    Check for next minimum TE path.
   ii.   GO TO 8.
```

Fig. 1 Algorithm for proposed system

Here, SN = Source Node, DN = Destination Node, RTE = Required Transmission Energy to transmit a packet from one node to next node on the path, NN = Neighbor Node, HM = Hello Message, and RE = Residual Energy.

4 Implementation

In the implementation of proposed routing scheme and existing scheme, we consider 16 primary user nodes and 24 secondary users. Transmission range of each node is 400 m. PU nodes are fixed and SU can wander in a region of 1500 m * 1500 m with random velocity and direction. Setup parameters used for simulation are shown in Table 3.

4.1 *Setup Parameters*

See Table 3.

4.2 *Snapshots*

Figure 2 shows the path generated by proposed routing, shortest path routing, and minimum transmission power routing. Node 22 is the source and node 31 is the destination. Three routing strategies were used, namely shortest path, Minimum Total Power Routing (MTPR), and optimal routing to establish a route from source

Table 3 Set up parameters

Region	1500 m * 1500 m
Transmission range	400
Nodes(SU)	24
Nodes (PU)	16
Position of SU	Random
Position of PUs	Fixed
Max velocity	15 m/sec
Pause time	0 s
Number of iteration	25
Source node	Chosen randomly from SU
Destination node	Chosen randomly from SU
Number of channels per user/node	1

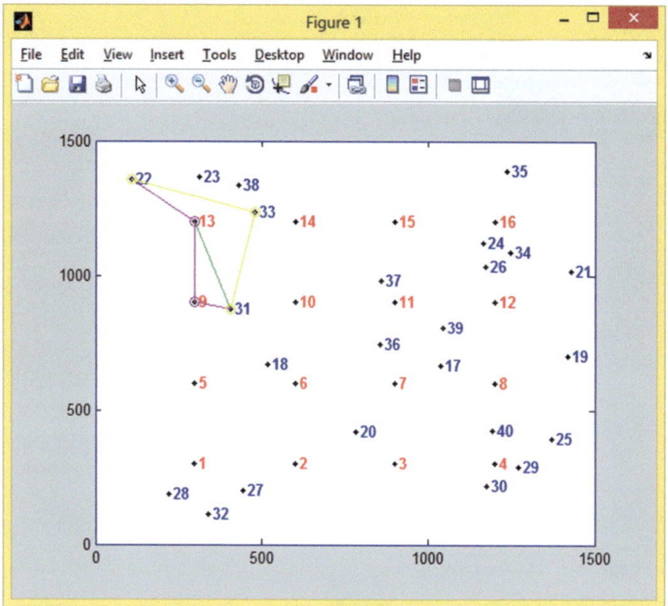

Fig. 2 Snap shot of simulation process, routing from (22) to (31)

to destination. The path from node (22) to node (31) shown in pink color shows the path formed using MTPR, in green using shortest path routing, and in yellow using optimal routing (see Fig. 2).

4.3 Results

Following observations were recorded for 25 iterations (Figures 3, 4, 5, and 6). Figure 3 shows the PDR values for all the three routing techniques. PDR value is quite high for optimum routing protocol.

Figure 4 shows average transmission power consumption for all the three approaches. The transmission power is required least in case of MTPR and may be equal to the optimum and maximum in case of shortest path routing approach, because a distant node requires more transmission power for transmission.

Figure 5 shows the variation of intermediate nodes in the path calculated. A path is considered to be more reliable if numbers of hops are lesser but as the same time nodes should not be much distant. Shortest path routing focuses on hops only, whereas MTPR focuses on less distant node, so hop count will be much higher. Hence, the hop count of optimal routing lies in between shortest path routing and MTPR.

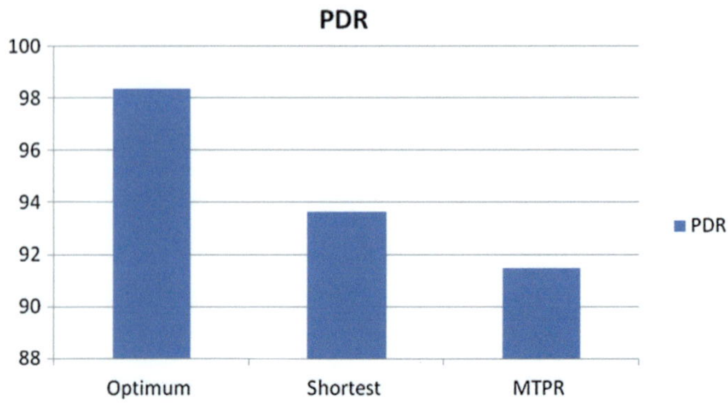

Fig. 3 Average PDR comparison

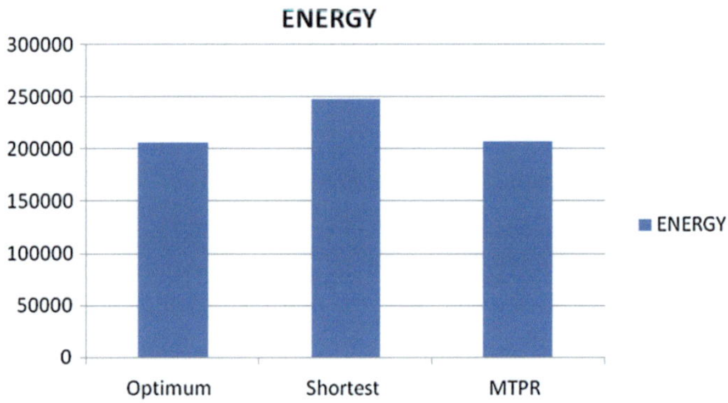

Fig. 4 Average energy comparison

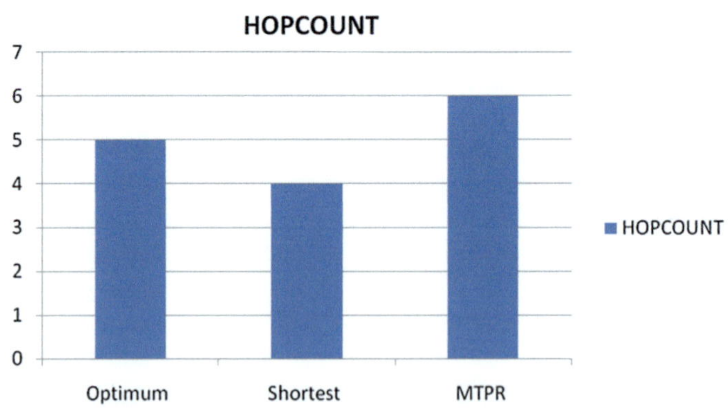

Fig. 5 Average hop count comparison

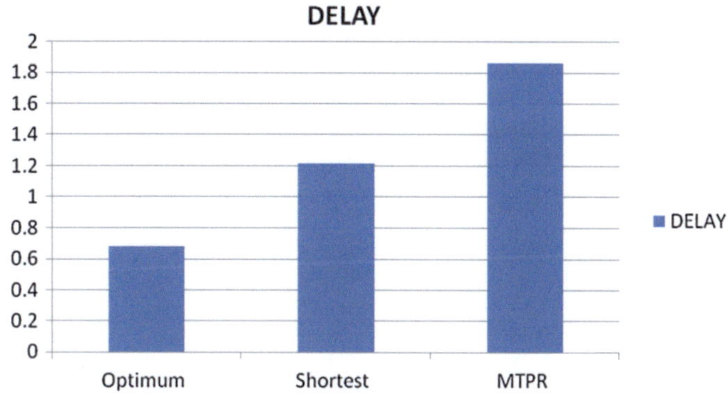

Fig. 6 Average delay comparison

Figure 6 shows the variation of end-to-end delay in different routing schemes. The value of end-to-end delay is much low for shortest path routing. But the delay will be highest in MTPR because there are maximum intermediate nodes. Delay in optimum routing mechanism is least because in shortest path routing delay may be contributed by unreliability of path.

5 Conclusion

Proposed routing protocol performs better than the traditionally used shortest path routing and MTPR. Because game theory analyzes the behavior of the nodes continuously, hence, the path obtained through optimized routing scheme will be more reliable. QoS parameters are optimized using proposed routing scheme. Game theory can be further used in CRN for dealing with various security issues.

References

1. Thomas RW, DaSilva LA, MacKenzie AB (2005) Cognitive networks., First IEEE international symposium on dynamic spectrum access networks, DySPAN, pp 352–360
2. Akyildiz IF, Lee WY, Vuran MC, Shantidev M (2006) NeXt generation/dynamic spectrum access/cognitive radio wireless networks-a survey. Comput Netw Int J Comput Telecommun Eng 50(13):2127–2159
3. Haykin S (2005) Cognitive radio: brain-empowered wireless communications. IEEE J Sel Areas Commun 23(2)
4. Yucek T, Arslan H (2009) A survey of spectrum sensing algorithms for cognitive radio applications IEEE Commun Surv Tutorials 2
5. Nasif AO, Mark BL (2008) Collaborative opportunistic spectrum access in the presence of multiple transmitters. IEEE global telecommunications conference 1–5

6. Setoodeh P, Haykin S (2009) Robust: transmit power control for cognitive radio. In: Proceedings of the IEEE, vol 97, pp 915–939
7. Ji Z, Ray KJ (2006) Belief-assisted pricing for dynamic spectrum allocation in wireless networks with selfish users. IEEE SECON. IEEE Press, Linthicum, pp 119–127
8. Wang W, Liu X (2005) List-coloring based channel allocation for open-spectrum wireless networks. In: IEEE Fall VTC 2005. IEEE Press, Dallas, pp 690–694
9. Neumann JV, Morgenstern O (2004) Theory of games and economic behavior. Princeton University Press
10. Li Y-B, Yang R, Yeo F (2010) Non-cooperative spectrum allocation based on game theory in cognitive radio networks. IEEE Fifth international conference on bio-inspired computing: theories and applications (BIC-TA), pp 1134–1137
11. Von Ahn L (2008) Preliminaries of game theory. Retrieved 2008
12. Bhattarai A, Komolkiti P, Aswakul C (2013) Improved bandwidth allocation in cognitive radio Networks based on game theory. IEEE tenth international conference on electrical engineering/electronics, computer, telecommunications and information technology (ECTICON), pp 1–6
13. Akyildiz IF, Lee W-Y, Chowdhury KR (2009) CRAHNs: cognitive radio ad hoc networks, Ad Hoc networks, vol. 7, pp 810–836
14. Liu Y, Peng QC, Shao HZ, Chen X, Wang L (2010) Power control algorithm based on game theory in cognitive radio networks. In: International conference on apperceiving computing and intelligence analysis (ICACIA), pp 164–168
15. Wang W, Cui Y, Peng T, Wang W (2007) Non cooperative power control game with exponential pricing for cognitive radio network. In: Proceedings 56th IEEE INT. Vehicular technology conference VTC2007-Spring, pp 3125–3129
16. Shiyin L, Mengyun L, Qiong L (2008) Dynamic power control algorithm and simulation in cognitive radio system. In: IET international communication conference on wireless mobile and computing CCWMC, pp 188–191
17. Liu Y, Peng Q-C, Shao H-Z, Chen X-F, Wang L (2010) Power control algorithm based on game theory in cognitive radio networks. In: International conference on apperceiving computing and intelligence analysis
18. Canales M, G´allego JR, Arag´on RC (2011) Distributed channel allocation and power control in cognitive radio networks using game theory. In: Vehicular technology conference IEEE
19. Qu Z, Qin Z, Wang J, Luo L, Wei Z (2010) A cooperative game theory approach to resource allocation in cognitive radio networks. In: 2nd IEEE conference on information management and engineering
20. Del Re E, Gorni G, Ronga L, Suffritti R (2009) A power allocation strategy using game theory in cognitive radio networks. In: International conference on game theory for networks

Study of Effect of Strain, Quantum Well Width, and Temperature on Optical Gain in Nano-Heterostructures

Swati Jha and Ashok Sihag

Abstract In the work discussed here, we evaluate the effect of change in strain, quantum well width variation, and temperature on the optical gain of two SQW (Single quantum well) nano-heterostructures. Both the heterostructures are SCH (Separate confinement heterostructures) with STIN (Step Index) profile. We have taken a quaternary semiconductor $Al_{0.15}In_{0.22}Ga_{0.63}As/GaAs$ and compared it with a ternary semiconductor heterostructure $In_{0.45}Ga_{0.55}As/InP$. This paper is an effort to compare the effect of change of strain on the optical gain of the two heterostructures. It also analyzes the behavior of quantum well width and temperature on the gain.

Keywords Heterostructures · Material gain · Strain · Temperature
Quantum well width

1 Introduction

InGaAlAs/InP heterostructures, in recent researches as in [1–3], have got special attention because of its lasing wavelength of 1.55 μm which dramatically coincides with the wavelength at which loss is minimum in optical fibers. In this work, we present the compositional details of the two different nano-heterostructures and investigate their comportment in terms of optical gain as against their lasing wavelengths. The effect of compressive and tensile strains is gauged on the two heterostructures along with the changes when the step index profiled well width of single quantum well is varied in steps on the material (optical) gain of the lasing herterostructures. Also, the influence of temperature variation is evaluated to figure out its impact on the gain.

S. Jha (✉) · A. Sihag
BML Munjal University, Gurgaon, India
e-mail: swati.jha@bml.edu.in

A. Sihag
e-mail: ashok.suhag@bml.edu.in

A. Mishra et al. (eds.), *Silicon Photonics & High Performance Computing*,
Advances in Intelligent Systems and Computing 718,
https://doi.org/10.1007/978-981-10-7656-5_6

2 Heterostructures Compositional Details

The compositional details of the STIN SCH SQW quaternary semiconductor laser of $Al_{0.15}In_{0.22}Ga_{0.63}As$ on GaAs are presented in Table 1, while Table 2 envisages representation of the second ternary semiconductor lasing heterostructure, namely $In_{0.45}Ga_{0.55}As$ on InP substrate.

In the work presented in this paper, the researchers evaluate the effects of changing strain, well width modulation, and temperature variation on the optical gain. A comparative analysis of the two SQW heterostructures is also carried out to evaluate and understand their usage and effectiveness as lasing heterostructures. The differences studied here suggest the variability of their usage and suggest the predictability of their applications.

3 Effect of Strain

It is well-known fact that compressive strain splits the first light-hole and heavy-hole subbands [4]. When the QWs are strained compressively, the differential gain in lasers improves because of the reduction in density of states (DOS) of holes (which gets increased because of heavy-hole and light-hole subband mixing). When the QWs are heavily compressively strained, hole confinement increases leading to improved threshold, injection efficiency, and ultimately the gain.

Table 1 $Al_{0.15}In_{0.22}Ga_{0.63}As$ quantum well lasing heterostructure

Layer number	Name of the layer	Material used	Width (nm)	Conduction band offset (eV)	Valence band offset (eV)	Lattice constant (Å)
1	Cladding	$Al_{0.61}Ga_{0.39}As$	10	0.6084067	−0.2863090	5.815376
2	Barrier	$Al_{0.2}Ga_{0.8}As$	5	0.2405987	−0.1132229	5.970752
3	Quantum well	$Al_{0.15}In_{0.22}Ga_{0.63}As$	6	0.0964814	−0.048240	5.909987

Table 2 $In_{0.45}Ga_{0.55}As$ quantum well lasing heterostructure

Layer number	Name of the layer	Material used	Width (nm)	Conduction band offset (eV)	Valence band offset (eV)	Lattice constant (Å)
1	Cladding	$Ga_{0.48}In_{0.52}As$	10	0.4844044	−0.1883795	5.863952
2	Barrier	$Al_{0.29}Ga_{0.17}In_{0.54}As$	5	0.1062417	−0.0413162	5.873997
3	Quantum well	$In_{0.45}Ga_{0.55}As$	6	−0.0283436	0.0141718	5.876105

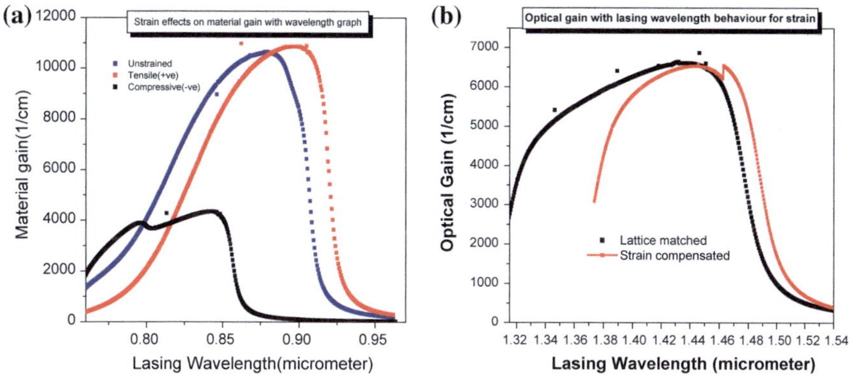

Fig. 1 Effect of strain on material gain as a function of lasing wavelength for **a** $Al_{0.15}In_{0.22}Ga_{0.63}As/GaAs$ and **b** $In_{0.45}Ga_{0.55}As/InP$

However, this methodology works well only till 1% of the strain [5]. Beyond that, strains hardly have any effect on DOS because at higher values of strain, heavy-hole and light-hole subbands are already separated in energy. In Fig. 1a and b, we plot the strain effects on optical gain when they are plotted with respect to the lasing wavelength for the heterostructures and observe the effect of unstrained/ lattice-matched and strain-compensated condition. From Fig. 1a it is observed that in $Al_{0.15}In_{0.22}Ga_{0.63}As/GaAs$ case, tensile strain (1.6×10^{-2}) gives the maximum gain; however, the lattice-matched condition is comparable. Nevertheless, if we consider the lattice-matched (unstrained) situation with the condition on application of strain (5.123×10^{-3}) for $In_{0.45}Ga_{0.55}As/InP$ in Fig. 1b, we observe a transition in the trend after the lasing wavelength of 1.449 μm at which the maximum gain of 6521.09 cm^{-1} in both cases is attained. We also find out that the peak material gain has a narrow spectrum in the strain compensated case.

4 Quantum Well Width Alteration Effects

Controlling the width of the quantum wells, the electron and hole wave functions can be transformed. This attribute leads to the modification of material parameters and not only improves the laser characteristics but also introduces new concepts to semiconductor optical devices. Here, in Fig. 2a, we investigate the effect of changing the well width from 60, 50, 40, and finally 30 Å on optical gain. Statistical data shows a staggering gain of 13,903.19 cm^{-1} at a lasing wavelength of 0.77 μm when the quantum well width is 40 Å. On increasing the well width from 30 Å to 50 Å and then to 60 Å, we observe a decrement in the gain values as well as the broadening of the gain spectra which may be attributed to the spectral hole burning.

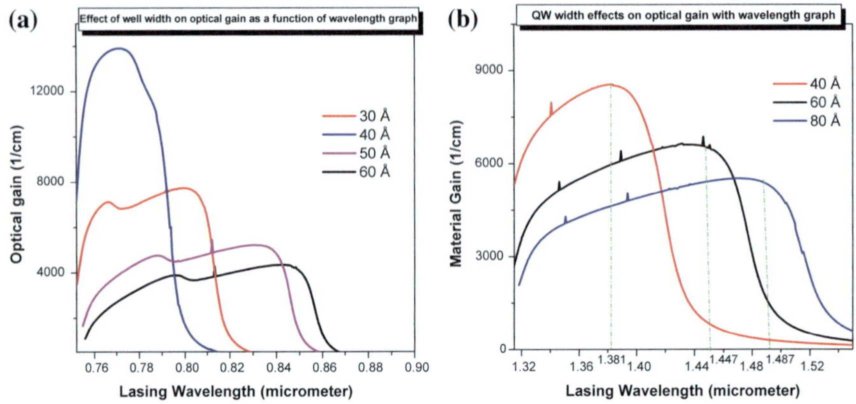

Fig. 2 Effect of QW width on material gain with lasing wavelength graph for **a** $Al_{0.15}In_{0.22}Ga_{0.63}As/GaAs$ and **b** $In_{0.45}Ga_{0.55}As/InP$

Investigating the effect of quantum well width variations from 60 to 40 Å, and finally to 80 Å on the optical gain for $In_{0.45}Ga_{0.55}As/InP$ illustrates that the maximum gain of 8550.24 cm^{-1} is achieved at a lasing wavelength of 1.38 μm for 40 Å. For 60 Å, a maximum gain of 6566.08 cm^{-1} is achieved at a lasing wavelength of 1.44 μm while for 80 Å, a maximum gain of 5373.29 cm^{-1} is achieved at a lasing wavelength of 1.48 μm as evident from Fig. 2b. It is indeed noteworthy that in both cases, at 40 Å, we have the maximum optical gain and also the gain spectrum is sharpest which indicates that it is the most suitable well width as far as gain is concerned because it offers maximum confinement. This observation reiterates the fact that maintaining a proper well width is very crucial for a better lasing characteristic.

5 Role of Temperature

Temperature has a big role to play in the lasing phenomena as it affects the injection efficiency, threshold current, and surely the gain. Hence, we now plot the effects of change in temperature on optical gain when it is changed from 100 to 300 K in steps of 100 K in Fig. 3a for $Al_{0.15}In_{0.22}Ga_{0.63}As/GaAs$.

Data from the graph indicates that at the same lasing wavelength of 0.84 μm we attain a maximum gain of 7494.40 cm^{-1} at 100 K, whereas at 200 K the gain is only 5096.12 cm^{-1} and the least value of gain (4321.95 cm^{-1}) is observed at 300 K. Exploring the effects of change in temperature on the material gain when it is varied from 100 to 300 K and finally 400 K in Fig. 3b for $In_{0.45}Ga_{0.55}As/InP$, we observe a regular trend of decrease in the material gain as we keep on increasing the temperature. This behavior of gain with temperature is similar in the two heterostructures.

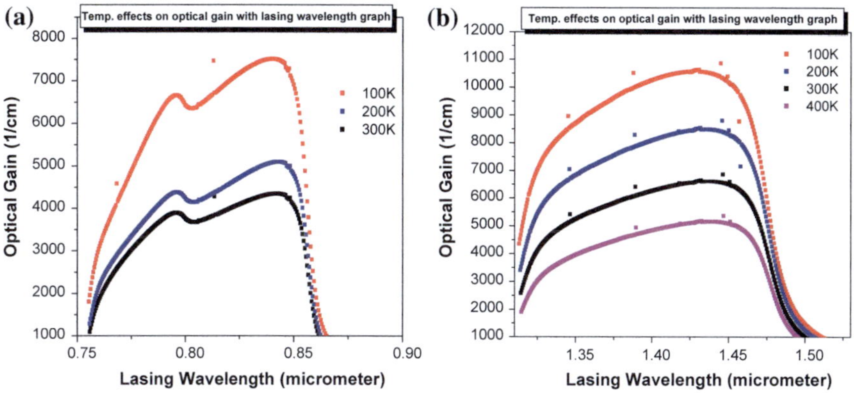

Fig. 3 Effect of temperature on material gain with lasing wavelength graph for **a** $Al_{0.15}In_{0.22}Ga_{0.63}As/GaAs$ and **b** $In_{0.45}Ga_{0.55}As/InP$

6 Conclusion

On comparing the two STIN SCH SQW nano-heterostructures, we observe that the effect of strain, in $Al_{0.15}In_{0.22}Ga_{0.63}As/GaAs$ case tensile strain (1.6×10^{-2}), gives the maximum gain; however, the lattice-matched condition is comparable. Nevertheless, if we consider the lattice-matched (unstrained) situation with the condition on application of strain (5.123×10^{-3}) for $In_{0.45}Ga_{0.55}As/InP$, we observe a transition in the trend after the lasing wavelength of 1.449 μm at which the maximum gain of 6521.09 cm^{-1} in both cases is attained. It is also observed that the peak material gain has a narrow spectrum in the strain compensated case. Investigating the well width variations yield, 40 Å, as the most suitable well width as far as gain is concerned for both cases, we not only have the maximum optical gain at this well width but also the gain spectrum is sharpest which indicates that it offers maximum confinement. This observation reiterates the fact that maintaining a proper well width is very crucial for a better lasing characteristic. On exploring the effects of changes in temperature on the material gain, we observe a regular trend of decrease in the material gain as we keep on increasing the temperature. This behavior of gain with temperature is similar in the two heterostructures.

References

1. Alvi PA, Lal P, Dalela S, Siddiqui MJ (2012) An extensive study on simple and GRIN SCH-based In0.71Ga0.21Al0.08As/InP lasing heterostructures. Phys Scr 85:035402
2. Alvi PA, Lal P, Yadav R, Dixit S, Dalela S (2013) Modal gain characteristics of GRIN-InGaAlAs/InP lasing nano-heterostructures. Superlattices Microstructures 61:1–12
3. Lal P, Yadav R, Rahman F, Alvi PA (2013) Carrier Induced Gain Simulation of InGaAlAs/InP Nano-Heterostructure. Adv Sci Eng Med 5:918–925

4. Vahala KJ, Newkirk MA (1989) Parasitic-free modulation of semiconductor lasers. IEEE J Quantum Electron 25:1393–1398
5. Ishikawa M, Nagarajan R, Fukushima T, Wasserbauer J, Bowers J (1992) Long wavelength high-speed semiconductor laser with carrier transport effects. IEEE J Quantum Electron 28:2230–2241

A Survey on Scheduling Algorithms for Parallel and Distributed Systems

Rinki Tyagi and Santosh Kumar Gupta

Abstract Task scheduling plays a vital role in distributed computing. It enhances the performance of the system as it minimizes the overall execution time and reduces overhead problems like communication delay by allocating suitable task to appropriate processor. Different scheduling techniques are discussed in this paper which are employed for task scheduling. Taxonomy of hierarchical classification is discussed for concurrent system and further several task scheduling algorithms are described on the basis of dependency and approach used such as static or dynamic for low throughput and efficient performance.

Keywords Task scheduling · Distributed computing · Real-Time scheduling

1 Introduction

Due to advancement of hardware and software technologies, development of parallel and distributed system for database, real-time applications are also increased. But it leads to a problem of scheduling different tasks on various processing elements in such a way that performance metrics, such as execution time, resource utilization, throughput, and response time, should be satisfied [10]. Scheduling is a type of resource and task allocation problem [11]. Scheduling can be defined as "A set of tasks T can be executed on P processors by some optimization criteria C" [11]. The goal of scheduling is to allocate different tasks to processors with the aim of increasing execution speed, reducing the runtime of tasks, minimizing communication delay, communication cost, and priority problem. In distributed scheduling, whole task is divided into sub-tasks and assigned to

R. Tyagi (✉) · S. K. Gupta
Department of Computer Science and Engineering, Krishna Institute of Engineering and Technology, Ghaziabad 201206, India
e-mail: rinkitg61@gmail.com

S. K. Gupta
e-mail: santoshg25@gmail.com

A. Mishra et al. (eds.), *Silicon Photonics & High Performance Computing*, Advances in Intelligent Systems and Computing 718,
https://doi.org/10.1007/978-981-10-7656-5_7

several processors, thus they execute more quickly as compared to single processor. While scheduling tasks, prespecified precedence is obeyed among different tasks [12]. The good scheduler should be General Purpose: it is applicable for all types of applications with different attributes to execute them in an efficient way; Efficient: it enhances the overall performance of systems and reduced overhead problems; Fair: It should be fair. For example, when there are many tasks to execute, scheduler should maintain load through load balance; Transparent: result should not be affected by local or remote site executions; and Dynamic: A good scheduler should respond to local changes and it should fully exploit all resources available to it.

Before describing the approaches, techniques, and algorithms that are used for task scheduling, the role of scheduler and how scheduling of tasks is to be done on different workstations or processors is discussed. There are two inter-dependent steps of task scheduling: one is allocating the tasks to processors (space sharing) and another is to schedule it with time (time sharing). When job is assigned to a system, the complete information is noted by the system, i.e., CPU load, memory size, etc. Meanwhile, system also maintains a status table of different workstations over which job is to be scheduled. After this, job is divided into several components and assigned to different workstations on the basis of the status table information. Synchronization process is needed when tasks are executing on different workstations. Besides this, mechanism of process migration is also introduced during task scheduling. For example, if any processor is highly loaded, the job migrates lightly loaded processor to improve the overall performance of system [12].

2 Related Work

The concept of "Grid Computing" in distributed system is used to perform users tasks online at any place and at any time [1]. But it leads to a problem of uncertainty in scheduling overhead and response time during continuous task arrival and their execution process. To overcome this problem, MDP (Markov Decision Process) is introduced where it allocates the task arrival and execution pattern without uncertainty. In [2], author argued the most general problem of process distribution. To solve this, a modified version of AO* algorithm using statistical data as a heuristic function is used to find those processors which can execute the processes in most efficient way. In [3], author has proposed a general technique to design and implement priority-based resource scheduling in flow graph-based DSPS with priority metadata. Experimental results show the effectiveness of this approach. In [4], author discussed the problem of task scheduling for multi-core CPU and to solve this multistep, scheduling algorithm is proposed. A clustering algorithm that is based on SCAN to find clusters in a network in order to find parallelism is used to decrease the computation of scheduling. In [5], an adaptive distributed scheduling algorithm is introduced for multi-place parallel computation. A combination of remote spawns and remote work steals is used to reduce the overhead problems in scheduling and helps to maintain load balance with maximal affinity. In [6], the author addresses the problem of load balancing optimization. As the load is distributed randomly to all processors in the distributed system without any fixed affinity, the goal is to minimize the computation time while distributing the load with the limited communication delay. For each load transfer, sending and receiving sites are maintained to obtain optimal delivery and load transmission. In [7], author proposed long-term CPU load prediction method, namely process search method also called runtime prediction-based method which predicts long-term CPU load more accurately than the conventional methods. A prediction module selection using neural network is proposed that selects an appropriate prediction method according to the change in the state of CPU load and shows improvement in prediction accuracy. In [8], author addressed the problem of producing the optimal schedule using genetic algorithm which minimizes peak load and communication cost and maximizes load balancing and average CPU utilization. In [9], author introduced an integrated BOA approach to overcome the efficiency problem while solving the NP scheduling problems. In [35], a taxonomy of load sharing is introduced that includes source initiative and server initiative approaches for evaluation of performance.

3 Issues in Multiprocessor Scheduling

There are many issues during scheduling for multiprocessor system [11]. First is distinction between Policy and Mechanisms. Mechanism is the ability to do any action while Policy enables to do with that mechanism. For example, automobiles have the ability to travel with the speed of 160 km/h (mechanism), while legal speed is set below 160 km/h (Policy). Second, distinguish between distributed and parallel system. If the communication between symmetric multiprocessors is through shared memory, it is parallel but if communication among network of workstations, it is distributed. Next is the distinction between types of scheduling. One is local scheduling and another is global. Global scheduling is done before local scheduling [11] although migration changes the global mapping when task moves to a new processor. During migration, the system stops the tasks, saves its state, and transfers that state to new processor and then restarts the task. Due to migration, several overheads occur. Local scheduling decides which an appropriate set of tasks at a processor executes next on that processor. *Then*, there are two choices at that time of task allocation. Either several processors are assigned to a single job or several tasks are assigned to a single processor. Few processors assigned to a single job for execution; then, it is called as *space sharing* and it is a function of global scheduling. *Time sharing* is a function of local scheduling. In this, several tasks are assigned to a single processor for its execution.

4 Scheduling Techniques

In Fig. 1, scheduling techniques are categorized into two types: local and co-scheduling. Local scheduling consists of predictive that easily adopts new architectures and proportional sharing that executes at a uniform rate. Three co-schedulings are discussed in which gang co-scheduling is simple than other two [12].

4.1 Local Scheduling

In local scheduling, an individual site schedules the processes assigned to it to improve the overall performance [12]. Local scheduler requires global information for maximizing the performance of system. Many new scheduling techniques are developed such as proportional sharing schedule approach and predictive scheduling. A module that has advance reservation capability is discussed that possessed local scheduling [13]. Local scheduling is used in opportunistic routing [14], in which wireless topology is broken down into several sub-graphs and end-to-end transmission of different forwarders is done that proved more efficient in wireless network compared to traditional routing.

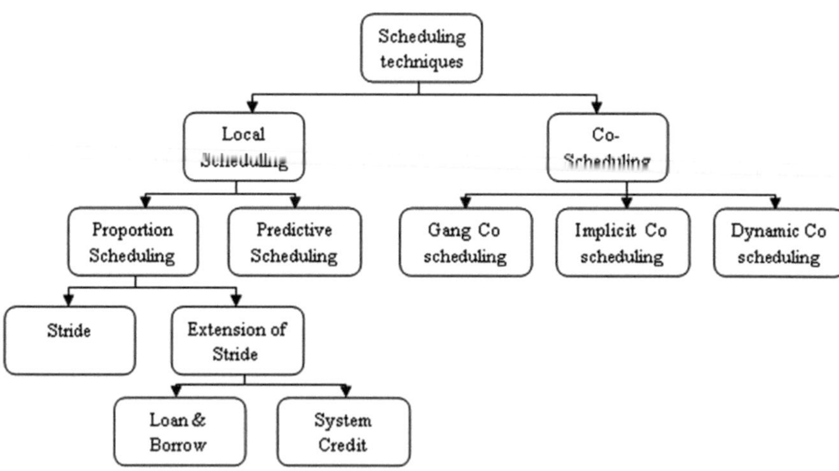

Fig. 1 Scheduling techniques

4.1.1 Proportional Sharing Schedule

To solve the problem incurred during traditional priority-based schedulers which take long time to allocate processors, a scheduler that allocates jobs to all processors in a fair manner is required [12]. In this, resource consumption is proportional to all jobs that are allocated. In [15], a notion of "Pessimism" is introduced in proportional sharing for improving performance as well as overcome error problem and meet the deadlines of various real-time applications. Through proportional sharing, scheduling is done at a uniform rate [16]. It is also beneficial for both real and non-real-time processing.

Stride Scheduling

It is an example of proportional sharing that shows how allocation of jobs and consumption of resources is done in fair manner to a single processor when many users have to execute their tasks. In this scheduling, users hold a number of tickets that are in a proportion of resources of competing users and have a time interval called stride which is inversely proportional to allocation of tickets that helps to decide how fast it comes in a usable state [12]. Also, a pass is associated with each user and the user with minimum pass is scheduled in that time interval and incremented by job stride. Its evaluation can be done by two ways: one is by using simulations and another is by implementing prototypes for Linux Kernel [17].

Extension to Stride Scheduling

Stride scheduling is only used for CPU-bound jobs, not for interactive and I/O intensive jobs [12]. In I/O intensive jobs, there is a need to improve response time and throughput rather than focusing on resources for competing users. For this, extension of stride scheduling is used. It uses two approaches: one is loan and borrow and another is system credits. Loan and borrow: In this approach, many exhausted tickets are distributed among the users. And, if any user wants to exit from system for a while, then another user borrows its ticket; otherwise, it would be an inactive ticket. System credits: It is an approximated approach and does not have any type of overhead and is easy to implement [18].

4.1.2 Predictive Scheduling

Predictive scheduling provides adaptively, intelligence and proactive type of scheduling to the system so that it can perform well in any type of environment. It can be easily embedded in new type of architectures. It is divided into three major components: H-Cell, S-Cell, and allocator [12]. In [19], a scheduling problem is investigated for minimizing completion time through random machine breakdowns.

Through predictive scheduling, preferences of users are also considered [20] and it is done through fuzzy techniques that not only use imprecise information but also views of users are to be considered.

4.2 Co-scheduling

Co-scheduling introduced in [21] is used for scheduling the interactive activities such that all jobs executed simultaneously on their workstations. In [22], flexible co-scheduling is introduced that addresses the problem of internal and external fragmentations. In this, scheduling is based on synchronization among processors and also on load balancing requirements. Co-scheduling was used for proper utilization of resources and covers all problems identified in multi-core system [23]. Several key issues are discussed in co-scheduling algorithm for clusters [39]. For this, co-ordinate scheduling in time-sharing clusters is done through a genetic framework.

4.2.1 Gang Co-scheduling

In gang co-scheduling, a job is referred as gang and its members are referred as gang members. They are allocated to a class whose one processor is assigned by one gang member and executed in parallel. A local scheduler exists in gang co-scheduling which has its own policies. When timestamps finish, it precepts all job members and assigns another job to that class [12]. The main disadvantage of this scheduling is its centralized control. It creates bottleneck when load is heavy. In [24], gang co-scheduling technique is combined with backfilling for addressing the problem of space sharing in scheduling. In [37], flexible co-scheduling is done through gang co-scheduling by reducing fragmentation problem and improves efficiency by providing the processor to each job through preprocess-based classification.

4.2.2 Implicit Co-scheduling

Implicit co-scheduling is a type of time-sharing communication process. It has a local scheduler that schedules the processes individually. It makes individual decisions while executing the job members rather following centralized policy and deals with the problem of gang scheduling [12]. It uses communication and synchronization within an application. It schedules both fine- and coarse-grained parallel application [25].

4.2.3 Dynamic Co-scheduling

Dynamic co-scheduling is used to make decisions on the arrival of messages. A schedule is made when a message is arrived to any process and no need for explicit identification to specify the process that needs co-scheduling. Dynamic co-scheduling reduces response time up to 20% over implicit [26] and is more robust and effective.

5 Taxonomy of Scheduling Algorithms

In Fig. 2, hierarchal classification of different scheduling algorithms is given. This classification can be used for categorizing different types of strategies that are used for allocating tasks to different processors and also for categorizing resource management system especially process management system.

Local versus Global: Local scheduling decides the appropriate set of tasks to execute next on the processor. Global scheduling is done before local scheduling and is used to allocate the tasks within the systems [27]. A new concurrency control criterion is proposed for local and flexible transactions execution through global scheduling in heterogeneous distributed environment [36].

Static versus Dynamic: Static algorithms are used for scheduling when information available at compile time. On the other hand, dynamic algorithm takes all these factors into account during execution time [27].

Optimal versus Sub-optimal: When information about the states of each processor and the types of resource needs is known, the algorithm is called as optimal otherwise sub-optimal.

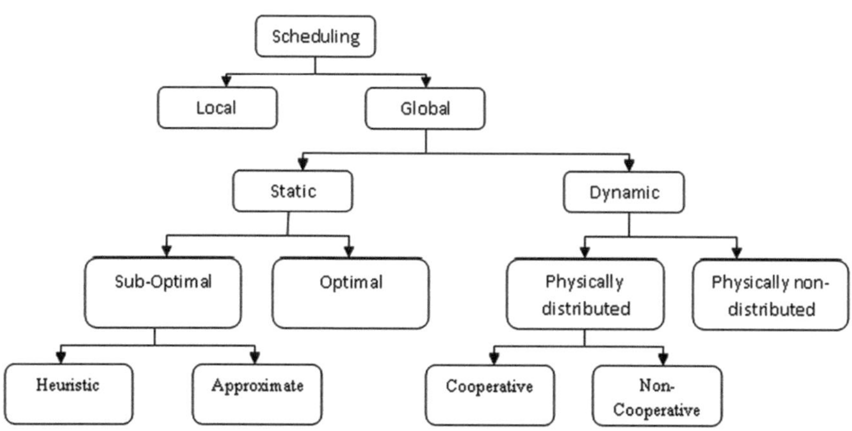

Fig. 2 Taxonomy of scheduling algorithms

Distributed versus Non-distributed: Physically non-distributed algorithms are centralized. A single processor decides task allocation. On the other hand, in physically distributed algorithm, all processors decide task placement [11].

Approximate versus Heuristic: Heuristic algorithms use various guidelines for scheduling such as allocate jobs to a processor with heavy inter-task communication, while approximate algorithms use same method that is used by optimal solutions but within a accepted range [27].

Cooperative versus Non-cooperative: In non-cooperative algorithms, each processor is independent of making choices from other processors for scheduling while in cooperative, all processors co-ordinate with each other to achieve a goal [28].

6 Task Scheduling Algorithms

While performing task scheduling, the performance of algorithms is affected by choosing various strategies. Different task scheduling algorithms are shown in Fig. 3.

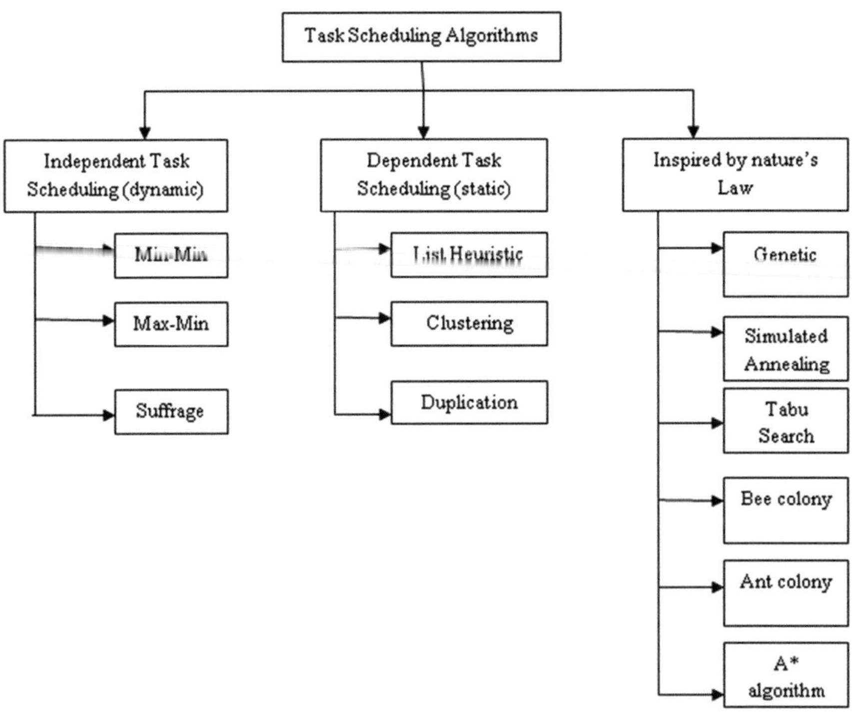

Fig. 3 Task scheduling algorithm

6.1 Independent Task Scheduling

This is used for dynamic scheduling. Some of the heuristic-based static algorithms also use this for execution of cost estimates [28]. *Max-Min* is somehow similar to *Min-Min* algorithm. In this, a set of M tasks of low completion time are selected from a set U of unmapped tasks. Then, the task of high execution time from M set is assigned to a processor. The newly mapped task is removed from set U and this process repeats until all the tasks of set U are not mapped. *Min-Min* is a type of scheduling which is applied for a set of tasks without dependencies onto a heterogeneous system. If there is a set of S tasks, then *Min-Min* selects a task with minimum execution time and it would be scheduled on that processor which has minimum completion time. It is one of the fastest algorithm and easy to implement. *Suffrage* is based on suffrage value and is calculated as the difference between its first best completion time and its second best completion time.

6.2 Dependent Task Scheduling

In [28], DAG (Direct Acyclic Graphs) is used to denote task precedence graph where node represents graphs and edge represents precedence order. The main objective of dependent task scheduling is to minimize the make-span time. But some problems are NP-complete which does not produce optimal results. It has three types of algorithms.

(i) *List Heuristic Algorithm*: It is based on the idea that some priorities are assigned to tasks and then repeatedly executes the following two steps until whole graph is not scheduled. First, remove the first node from list of scheduling. Then, assign the node to a processor that allows the earliest start time. Table 1 shows different algorithms for list heuristic.

(ii) *Clustering Heuristic Algorithm*: Clustering is an efficient process for both parallel and distributed systems. It is used to minimize communication delay

Table 1 List heuristic algorithms

Type	Description
HEFT	It is Heterogeneous Earliest Finish Time algorithm. At each step, it selects a task that has higher upward rank and with maximum distance between current and exist nodes
Chaining	It does not allow duplication of tasks and distribute the tasks among all processors
HLFET	Highest Level First with Estimated Times algorithm is one of the simplest scheduling algorithms. It calculates b-level (bottom level), and then makes a ready list in descending order (contains only entry nodes). It schedules first node of ready list to a processor with earliest execution
FCP	It is Fast Critical Path algorithm and tries to reduce the complexity of algorithm

Table 2 Clustering heuristic algorithms

Task clustering phase	DSC: It is Dominant Sequence clustering algorithm. The scheduled DAG has a critical path called Dominant Sequence and used to distinguish it from scheduled clustered DAG critical path
Post-Clustering Phase	LB: Load Balancing is used to define the overall workload as sum of execution time of all tasks. It merges a pair of clusters C1 and C2 such that C1 is the cluster that has minimum workload and C2 is a cluster that minimum workload with communication edge to all clusters CTM: It is Communication Traffic Minimization and it is defined by using two terms C1 and C2. These are the sums of communication time of edges from C1 to C2 and from C2 to C1 RAND: It is random algorithm and make random pairs of clusters. First, assign a cluster with highest communication traffic. Then, it selects an unassigned cluster of highest communication traffic through assigned cluster. Then repeat step(b) until all processors are assigned

by clustering the same level of tasks into one cluster and assign resources to a single cluster. Table 2 shows that first phase is task clustering phase in which clusters are formed by portioning the tasks and same levels of tasks are clusters into one group. Second phase is post-clustering phase which is used to refine the cluster to get final task to resource map [28].

(iii) *Duplication Heuristic Algorithm*: It is basically used for homogenous system. For this, various duplication-based scheduling algorithms are introduced. The main logic behind this is to fully utilize the idle resources by duplicated tasks.

6.3 Scheduling Inspired by Nature's Law

Genetic algorithm, simulated annealing, Tabu search, etc., are scheduling methods that are inspired by nature's law. They all are used to solve standardized scheduling problems. A long runtime problem will occur when they are used in practical application. Then, optimization algorithms are used to overcome the problem of long runtime [28]. Genetic algorithm optimizes the total flow time and makes span in task scheduling and is more robust. A simulator package of grid is used for large problems and to evaluate performance of GA [29]. Tabu search is also an optimization algorithm that gives optimal or close to optimal results for different types of scheduling [30]. Tabu search is used for bi-objective optimization problems [31]. Simulated annealing is to solve optimization problems for separating continuous and batch chromatographic systems under various conditions such as gradient and isocratic [32]. It states that if temperature is lowered sufficiently, then the solid reached to optimal state, where temperature is completion time and change in temperature is task mapping. And if temperature is increased, then it accepts that "worse" state because it provides a way to escape from local search [28]. Ant colony optimization is best suited for TSP problems. Ants act as an agent and finds the best solution by parallel search. Many ants generate many cities and their corresponding paths, and best suited path (shortest path between source and destination) is selected [33]. Bee colony optimization is best suited for combinatorial optimization problems. When independent tasks are scheduled in grid environment, then it is a NP-hard problem. Many approaches are used for this but bee colony is more efficient as it reduces the finish time and delays during execution time [34]. In [38], different applications are discussed for BCO. A* algorithm is a type of best search algorithm for tree search. It initially starts with null solution and then expanded through partial solution for complete solution by allocating different tasks to suitable processors [28].

7 Analysis Work

Table 3 shows the different scheduling techniques, task scheduling algorithms, and approach that can be used for scheduling tasks.

Table 3 Analysis work

Techniques	Description	Approach	Algorithms	Advantage
Proportional share	Resource consumption is proportional for all jobs	Static	List heuristic	Does not allow duplication of tasks
			Clustering heuristic	Minimize communication delay
Predictive	Adaptive, intelligent, and proactive techniques that perform well in any type of environment		Duplication	Fully utilized the idle resources
Gang co-scheduling	Dynamically scalable and utilized time sharing and space sharing simultaneously	Dynamic	Min-Min	Fastest algorithm and easy to implement
Implicit co-scheduling	A local scheduler that schedules the processes individually rather centralized policy		Max-Min	Removes the penalties occurred through longer execution time
Dynamic co-scheduling	Used to make the decisions on the arrival of messages		Suffrage	Finds minimum completion time

8 Conclusion

In distributed environment, task scheduling plays a vital role. To improve, the performance of system scheduling of tasks can be done in an efficient manner on several processors. A good scheduler must be efficient, dynamic, transparent, general purpose, and fair. Different scheduling techniques are employed for both parallel and distributed systems that are used for proper utilization of resources and for improving system performance. Two scheduling techniques, i.e., local and co-scheduling, are described in which co-scheduling technique is better as it is used for the concurrent execution of the parallel systems and we conclude that gang co-scheduling is the simplest one among three. A hierarchal classification of scheduling algorithms is done and on this basis task scheduling algorithms are described. Further, a comparative analysis for different scheduling techniques has been done.

9 Future Work

Further, more comprehensive survey can be made on real-time scheduling and scheduling inspired by nature's law and these can be compared with other scheduling algorithms.

References

1. Zhao T, Zheng X (2014) Proactive scheduling in distributed computing—A reinforcement learning approach. J Parallel Distributed Computing, pp 2662–2672
2. Gyire T (1995) A distributed process scheduling algorithm based on statistical heuristic search. IEEE International Conference
3. Bellavista P (2014) Priority-based resource scheduling in distributed stream processing systems for big data applications. Utility and Cloud Computing (UCC). In: IEEE/ACM 7th International Conference on IEEE
4. Yamazaki H, Konishi K, Shin S, Sauada, K (2013) Multistep scheduling algorithm for parallel and distributed processing in heterogeneous systems with communication costs. Mathematical Problems in Engineering
5. Narang A, Srivastava A, Shyamasundar RK (2013) High performance adaptive distributed scheduling algorithm. In: Parallel and distributed processing symposium workshops & PhD Forum (IPDPSW), 27th international IEEE
6. Haddad E (1994) Real-time optimization of distributed load balancing. Proceedings of the second workshop on IEEE
7. Sugaya Y, Tatsum H, Kobayashi M, Aso H (2008) Long-Term CPU Load Prediction System for Scheduling of Distributed Processes and its Implementation. Advanced Information Networking and Applications, 22nd International Conference on. IEEE
8. Wang PC, Korfhage W (1995) Process scheduling using genetic algorithms. Parallel and distributed processing, proceedings seventh ieee symposium on IEEE

9. Qiang L, Xiao TY (2006) Cooperated Bayesian algorithm for distributed scheduling problem. Frontiers Electr Electron Eng China, pp 251–254
10. Shirazi BA, Hurson AR, Kavi KM (1995) Introduction to scheduling and load balancing in parallel and distributed system. Wiley-IEEE computer society press
11. Chapin SJ, Weissman JB (1996) Distriibuted and Multiprocessor scheduling. published In: ACM computing survey(CSUR), 28:233–235
12. Dongning L, Ho PJ, Liu B (2000) Scheduling in distributed system
13. Nakada H, Kishimoto M, Kudoh, T, Tanaka Y, Sekiguchi S, Takefusa A (2006) Design and implementation of a local scheduling system with advance reservation for co-allocation on the grid. In: Computer and information technology, sixth IEEE international conference
14. Li Y, Liu YA, Li L, Luo P (2009) Local scheduling scheme for opportunistic routing. In: Wireless networking conference IEEE, pp 1–6, (2009)
15. Regehr J.: Some guidelines for proportional share CPU scheduling in general purpose operating system. In: Work in progress of the 22nd IEEE Real -time system symposium (RTSS) (2001)
16. Stoica I, Wahab HA, Jeffay K, Baruan SK, Gehrke JE, Plaxton CG (1996) A proportional share resources allocation algorithm for real time, time shared systems. IEEE, pp 288–299
17. Gu W, Carl A, Weihl WE (1995) Stride scheduling: deterministic proportional share resource management. Massachuesetts Institute of Tech, laboratory for computer science
18. Koshy R (2014) Scheduling in distributed system: a survey and future perspective. Int J Adv Technol Eng Sci
19. Xing Z, Zhijon C, Yugeng X (2007) The applications of predictive scheduling algorithm for single machine problem. In: Control conference IEEE, 810–814
20. Sauer J, Chua TJ (2014) Fuzzy predictive and reactive scheduling in soft computing for business intelligence. Springer Berlin Heidelberg, pp 281–297
21. Gupta A, Taucker A, Urushibaras S (1995) The impact of OS scheduling policies and synchronisation methods on performance of parallel applications. In: SIGMETRICS perform evaluation review
22. Frachtenberg E, Feitelson DG, Petrini F, Fernandez J (2005) Adaptive parallel job scheduling with flexible coscheduling. IEEE, pp 1066–1077
23. Schonhers JH, Lutz B, Richling J (2012) Non-Intrusive co-scheduling for general purpose operating system. Springer Berlin Heidelberg, pp 66–77
24. Zhang Y, Franke H, Moreira JE, Sivasubramaniam A (2000) Improving parallel job scheduling by combining Gang scheduling and backfilling techniques. IEEE, pp 133–142
25. Anglano C (2000) A Comparative evaluation of implicit coscheduling strategies for network of workstations. IEEE, pp 221–228, 1 Aug–4 Aug
26. Sobalvarro PG, Scott P, Weihl EW, Andrew AC (1998) Dynamic coscheduling on workstations clusters in Job Scheduling Strategies for Parallel Processing. Springer Berlin Heidelberg, pp 231–256
27. Casavant TL, Kuhl JG (1988) A taxonomy of scheduling in general-purpose distributed computing systems. Software Engineering, IEEE Transactions on 14(2):141–154
28. Shahsavari M, et al (2004) Task scheduling policies in general distributed systems: a survey and possibilities
29. Carretero J, Xhafa F (2006) Use of genetic algorithm for scheduling jobs in large scale grid application. ISSN 1392–8619 UKIO Technologinis IR Econominis Vystymas, pp 11–17
30. Glover F (1990) Tabu search: a tutorial. pp 74–94
31. Xhafa F, Carretero J, Dorronsoro B, Alba E (2012) A tabu search algorithm for scheduling independent jobs in computational grids. Comput Inform 28:237–250
32. Kaczmarski K, Antos D (2006) Use of simulated annealing for optimization of chromatographic separations. Acta Chromatographica 17
33. Darquennes D (2005) Implementation and Applications of Ant Colony Algorithms. Facultées Universitaires Notre-Dame de la Paix, Namur Institute Informatique

34. Mousavinasab Z, Entezarii ME, Movaghar A (2011) A bee colony task scheduling algorithm in computational grids. Digital Information Processing and Communications, Springer, Berlin Heidelberg
35. Wang Y Load sharing in distributed system. IEEE
36. Zhang A, Noidine M, Bhargava B (2001) Global Scheduling for flexible transactions in heterogeneous distributed database systems. 13(3):439–450
37. Frachtenberg E, Feitelson DG, Petrini F, Fernandez I (2003) Adaptive Parallel job scheduling with flexible co-scheduling. Parallel and Distributed processing, 10 pp, IEEE
38. Karwan KS, Choudhary S, Sharma K Applications of artificial bee colony optimization techniques. pp 1660–1664, IEEE, (2015)
39. Agarwal S, Yoo AB, Choi GS, Nagar S (2003) Coordinated co-scheduling in time sharing through a genetic framework. pp 84–91, IEEE, (2003)

Analysis of On Chip Optical Source Vertical Cavity Surface Emitting Laser (VCSEL)

Sandeep Dahiya, Suresh Kumar and B. K. Kaushik

Abstract The major developments in semiconductor laser technology, i.e., Vertical Cavity Surface Emitting Lasers (VCSELs), really revolutionized the field of semiconductor lasers and play a pivotal role in every walk of life such as science and technology, research and development, consumer and industrial environment, medical, military, surveillance, telecommunication, and a host of other applications. Although the development of lasers is pertinent to each other because of dependence on Distributed Bragg Reflector (DBR) mirrors, the devices under reference are normally supposed to be the most appropriate semiconductor lasers for their evident plentiful significances and applications. In modern age, the importance of VCSELs is reflected in fact that they have become the second largest production among all types of semiconductor lasers due to intrinsic structure features of array formation, coherent emission with small beam divergence, large output power, low-threshold operations, high modulation bandwidths, etc., tuned by electrical and temperature variations. In the present investigation, the electrical and optical characteristics of the state-of-the-art long-wavelength VCSEL at 1310 nm emission is analyzed with different apertures such as 20 and 12 μm. The authors observed that if the oxide aperture of the same device is reduced 20 to 12 μm, it obtained the incremental in the carrier and photon density rates and subsequently reduces the emitted power, threshold current, and gain of the devices. The present communication discusses the history, present status, and an exposure of some state-of-the-art performances with optimized results of VCSELs.

S. Dahiya (✉)
Department of Electronics and Communication Engineering,
BPSMV, Khanpur-Kalan, Sonipat, India
e-mail: sandy_dahiya2001@yahoo.com

S. Kumar
Department of Electronics and Communication Engineering,
UIET, MDU, Rohtak, India
e-mail: skvashist_16@yahoo.com

B. K. Kaushik
Department of Electronics and Communication Engineering, IIT, Roorkee, India
e-mail: bkkaushikiitr@gmail.com

A. Mishra et al. (eds.), *Silicon Photonics & High Performance Computing*,
Advances in Intelligent Systems and Computing 718,
https://doi.org/10.1007/978-981-10-7656-5_8

Keywords VCSELs · Device layout · Significance and applications
Characterization and results

1 Introduction

Lasers are one of the most significant inventions developed during the mid-twentieth century and play a pivotal role in tremendous aspects of uses in electronics, computer hardware, science, and technology. The first semiconductor p–n junction laser was reported in 1962 [1] and constant developments have been achieved which appreciably increasing the performances of the lasers. But due to the lack of optical and electrical confinement amongst the semiconductor p–n junction lasers, the major advance was recognized for the first time in 1968–1969 with the development of the double-heterojunction laser allowing to reach room-temperature operation. Further, by the impact of optical attenuation and scattering effects in long-haul fibers, there was a dire need for high-intensity transmitters. In optics, the physics is determined by the nature of light and the data is modulated on to optical signals to increase the bandwidth at low power dissipation. The right candidate of optical source is VCSEL which can operate at 850–1550 nm wavelength with single-mode or multimode emission. VCSEL is a type of semiconductor laser with the cavity perpendicular to the wafer plane; its structure allows the emission of light or optical beam in vertical direction either to top (top-emitting devices) or to downward (bottom-emitting devices) direction.

The concept of the VCSEL was first proposed by the elite group of Professor K. Iga of Tokyo Institute of Technology in 1977 and the first working VCSEL devices were accordingly reported in a while by the same group [2]. The preliminary VCSELs with pulsed operation had metallic mirrors and operated at very high threshold currents app. 1000 mA at temperatures 77 K as reported by Soda et al. in 1979 for VCSELs emitting at 1310 nm with GaInAsP/InP for active region and 6 mA threshold device was reported by the same group in 1986 and later Chow et al. presented the fabrication and performance of infrared and visible VCSELs in 1997 [3, 4]. But in next decade, some improvements have been made by the same group of Professor Iga about the prospects of VCSEL devices in the GaInAsP/InP and AlGaAs/GaAs materials. Subsequently, the first room-temperature VCSEL came into existence with two dielectric mirrors [5], and in 1989 Jewell et al. presented 850 nm VCSELs with continuous wave operation which came into existence in 1991 [6]. VCSEL really shows the presence of advances in epitaxial growth techniques of molecular beam epitaxy (MBE) and metal organic chemical vapor deposition (MOCVD). To make the device companionable with an industrial fabrication, the researchers used double-fused GaAs/GaAlAs mirror for 1550 nm [7, 8] and 1300 nm wavelength [9, 10] or by the combination of one dielectric mirror and one wafer fusion to obtain the good results. Moreover, in 1989, Lee et al. presented 2 mA threshold QW VCSEL [11], Baba et al. demonstrated CW operation for 1310 nm device in 1993 [12], and

20 mW high power was reported by Grabherr et al. in 1996 for the VCSEL emitting at 960 nm [13]. The better results provided by the high thermal conductivity and reflectivity intrinsic to the GaAs-based material is widely optimistic for short wavelength, i.e., 850 and 980 nm wavelength VCSELs. Although edge-emitting traditional lasers are competent of watt-level output power, their asymmetric optical confinement causes very low coupling efficiency into optical fibers. VCSELs possess higher coupling efficiency into fibers due to low-divergence circular output beams, low-threshold currents continuous wave operation, etc. [3, 14, 15]. The techniques MBE and MOCVD significantly improved the characteristics of the mirrors due to the DBRs [16] providing high reflectivity and low losses at low-threshold currents. The rapid progress by Coldren's team in 1991 by introducing new gain materials InGaAs quantum wells (QW) for the cavity of VCSEL was undertaken [17]. Further, the introduction of Al oxide confinement in VCSELs' technology such as [18] has led to even lower threshold values [19] and high-frequency modulation [20]. The research was accelerated to plunge the device in commercial market in the late 90s to meet the third generation of development in 1999 [2].

VCSELs have been of enormous outcome over recent years with their remarkable features such as single-longitudinal-mode operation, circular output beams with low-divergence, low-threshold currents CW operation, ability to be modulated at very high frequencies, and a significance progress that has been made in many diverse current technology applications. The device is considered one of the most vital companions for the performance of the optical interconnects and facilitating ultra-parallel information broadcasted in lightwave and computer systems. Generally, there are two types of VCSELs in existence emitting at short-wavelength spectral range, i.e., wavelength in between 800 and 1000 nm and long-wavelength VCSELs emitting greater than 1000 nm. In past decade, almost all commercial VCSELs were designed for multimode emission but nowadays single-mode emission devices have captured half of the market volume [21–24].

The present investigation will provide an overview of historical path and present status to give an exposure of some state-of-the-art VCSELs performances with optimized results for long-wavelength 1310 nm emission. The manuscript is organized as follows: Sect. 2 presents the device structure and operation; in Sect. 3, the advantages and applications of devices are presented; and Sect. 4 is devoted to the characterization and results of VCSEL summarized the concluding remarks. The succeeding section of study will present the VCSEL structure and its operation.

2 Device Structure and Operation

Nowadays, significant research and development efforts have been reported by researchers and scientists toward VCSELs due to their remarkable features. The VCSEL structure is unique than the conventional edge-emitting lasers or stripe lasers because it emits light perpendicular to the surface of semiconductor wafer by making more features available on it. The device structure comprised of a top p-type DBR mirror, an active region (cavity), and a bottom n-type DBR mirror to provide the essential optical feedback.

The active layer thickness based on quantum wells or quantum dots is much shorter than cavity length formed by epitaxial layers. Since they gain path along the emission, i.e., in vertical direction, the cavity length of VCSEL is of few tens of nanometers, i.e., hundred times lesser than that of traditional lasers because very high reflectivity required for optical feedback. The transverse symmetry of the cavity causes a low-divergence angle circular output beam [25]. This optical confinement has been obtained through the current confinement schemes such as index guiding via selectively oxidized apertures [19], gain guiding via proton implantation [26], as well as two-dimensional photonic crystals for endless single-mode operation [27]. The properties of VCSEL make it advantageous over traditional edge-emitting lasers because reflectivity of mirror has to be nearly 99–99.9%, whereas it is merely 3% in case of edge-emitting lasers. The emission of light or power is from the top mirror of the device as demonstrated in Fig. 1 [2].

Figure 2 shows the device structure at 1310 nm wavelength emission with 20 um oxide aperture generated by simulation package. The DBR mirrors as shown in Fig. 3 comprised of alternating sequence of high and low refractive index layer material represented by n with quarter wavelength thickness, where λ is the free space wavelength [28]. Generally, to achieve the better reflectivity and electrical conductivity, more than 20 pairs of each top and bottom mirror are required. The

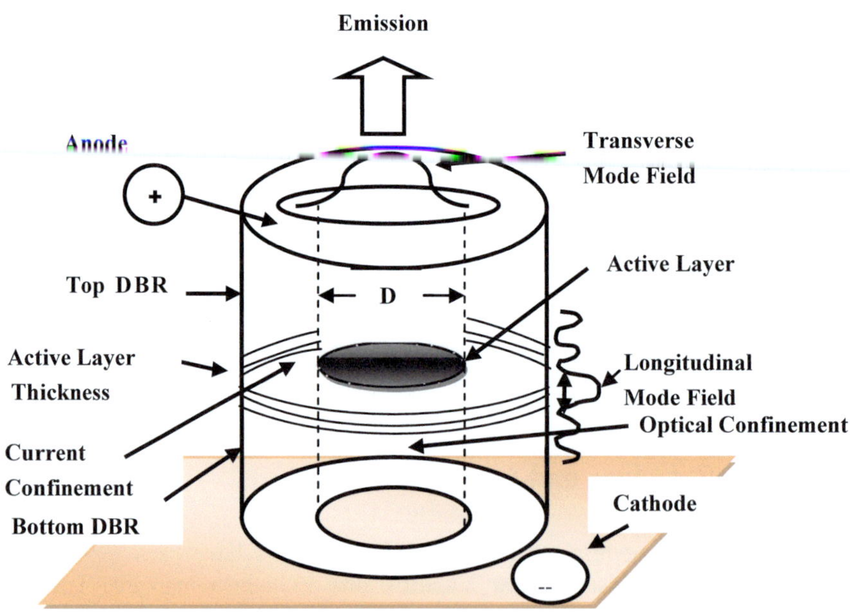

Fig. 1 Layout structure of vertical cavity surface emitting laser

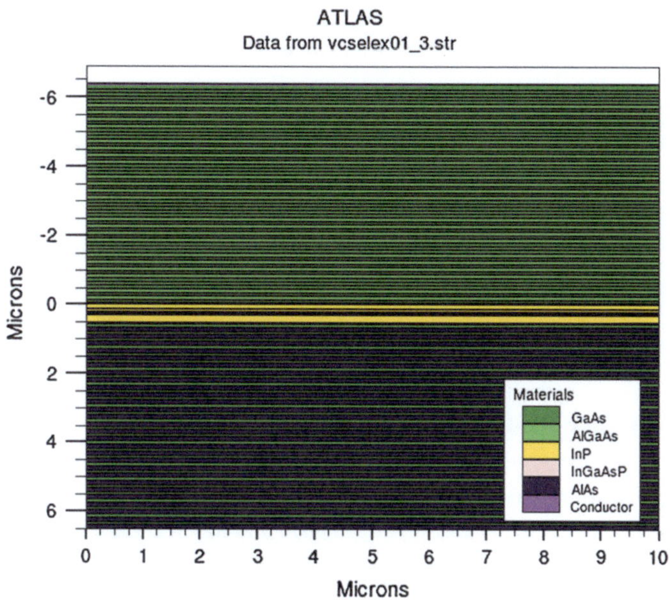

Fig. 2 VCSEL structure created by simulation package

Fig. 3 DBR structure

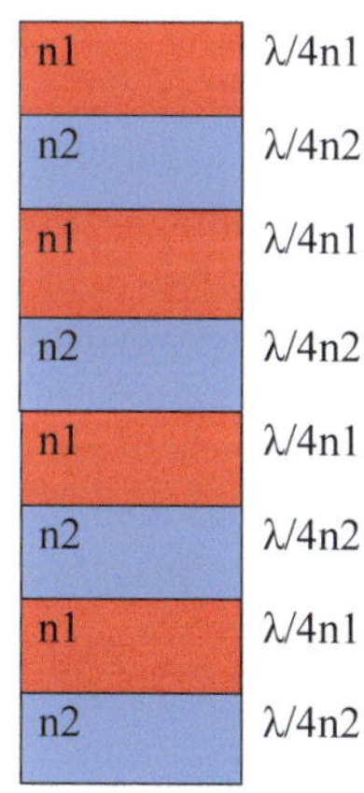

two top and bottom DBRs in the VCSEL are oppositely doped, i.e., n-type DBR and p-type DBR. The active region emitting light, generally including undoped single or numerous quantum wells, receives current through a current conducting structure which results in high optical power, low divergent angle, and circular symmetric beam in a superior quality [29–31]. Most common quantum well materials assisting direct transitions comprise GaAs for 850-nm-short-wavelength emission and GaInNAs for 1310-nm-long-wavelength emission [32–34].

Moreover, for integration in three-dimensional micro-optical systems, the vertical emission of light sources (VCSELs) is desirable. The structure of VCSEL is optimized for the lithographic fabrication of densely packed two-dimensional arrays. Therefore, VSELs are perfect light sources for three-dimensional micro-optic, where two-dimensional data are processed in parallel. Finally, the device layout structure and small mode size enable fabrication of extremely consistent and compactly packed lasers with nominal crosstalk for use in parallel optical interconnects [35–37]. The succeeding section will discuss a few among the numerous advantages of VCSELs including optical interconnects, parallel fiber optic systems, medical, data communication, signal processing, etc.

3 Significance and Applications of VCSEL

VCSELs have turn out to be long strides in the short duration of its existence due to its enormous progress of VCSELs performance and applications. Although the first lasing operation on VCSEL was experimental in 1979 in modern information and technology era, they have various advantages over edge-emitting lasers such as low-threshold operations, low cost which makes it commercially viable, single-mode operations, high-speed operations, efficient sources for coherent radiation [2, 21, 38, 39], temperature sensitivity of wavelength, high modulation bandwidth [40], low-divergence circular output beams, low power consumption, compactness [21, 41], high-speed data transfer [42], optical data interconnects [43], efficient coupling to optical fibers, two-dimensional array fabrication by entirely monolithic processes, packaging and heat sinking [44], optical information processing functions [21], pumping for solid-state lasers [45], device structure allowing wafer-level testing, thereby low-cost chip production [2], high power adaptation efficiency, i.e., more than 50%, suitable for optical storage and telecom appliances, due to which embedded active region lifetime is longer, index-guided structure with an oxide aperture and multiple quantum wells in active region, increase in the system throughput, probable sources for plastic fiber communication, free space optoelectronic processing [2, 38, 39], and a host of other significant applications of VCSELs. Long-wavelength VCSELs have also been used in some emerging areas such as Si-related technologies and sensor system [46, 47].

Apart from the above, VCSEL became useful for a diversity of consumer and industrial environment, medical, and military applications requiring high power and energy, such as medical and cosmetics, infrared illuminators for military and surveillance, pumping of solid-state lasers and fiber lasers and also applicable in biological tissue analysis, gas sensing, laser printers, biophotonic chips, etc. [33, 34, 48–50]. In past several years, the commercialization of VCSELs with above-discussed performances and applications make it a strong contender over conventional lasers.

This section deliberates on few significances and applications of VCSELs; the next section of the investigation reports the characterization and optimized results of VCSELs.

4 Characterization and Results

Optical interconnects are promising technology to fulfill the high-speed requirements of next generation. Optical interconnects exhibit a variety of advantages over copper and electrical interconnects such as higher bandwidth, low crosstalk between channels, and low power consumption. Moreover, it is vital to decrease the volume of active region (cavity) in order to reduce the threshold current in case of VCSELs. Ikechi et al. in 2013 demonstrated the performance comparison of the high-speed characteristics of VCSELs for the short wavelength (850 nm emission) and long wavelength (1310 nm emission) for top- and bottom-emitting surface for use in modern high-bandwidth fiber optic networks [51]. The VCSEL's exceptional characteristics such as parallel operation, high modulation rate, and circularly symmetric emitted beam make them very gorgeous light sources for the next generation of LAN, ethernet, and optical interconnects. The long-wavelength VCSELs are advantageous for parallel light wave structures to assemble speedy enhancement of information transmission capability in local area networks. Semiconductor VCSEL layouts are appropriate for exploring the characteristics of the spontaneous emission since the device structures are fabricated by MBE and MOCVD technologies. The reflectivity of the spontaneous emission can be simply deliberated by varying AlAs/GaAs pairs of DBR mirrors. The communication between excitonic energy and active region (cavity mode) can be analytically changed by employing thermal characteristics of VCSEL layout structures [52, 53].

In present analysis, the characteristics of 1310-nm-long-wavelength VCSEL of oxide aperture 20 μm is demonstrated. The VCSEL under reference utilize strain-compensated InGaAsP six MQ wells optical cavity sandwiched between GaAs/AlGaAs mirrors, those are double-fused on the InP space layers in active region at both sides. The top mirror consists of 30 pairs of materials GaAs and AlGaAs and is covered by GaAs layer to enhance the reflectivity with 28 pairs of bottom mirror of material GaAs and AlAs quarter-wave layers. The simulated waveforms obtained by commercial software tool for laser power, gain, photon density, optical wavelength, and mirror loss versus applied voltage are shown in Fig. 4. On analyzing the obtained results for 1310-nm-long-wavelength emission, the laser or emitted power achieved is 98.6 mW, gain is 30 cm^{-1}, photon density is $2.88 \times 10^6/cm^3$, and mirror loss is 30.7 W/cm with reflectivity of DBR mirrors greater than 99%. The amplification must be able to offset all the losses to obtain lasing operation to maintain the reflectivity of mirrors very high. The authors analyzed that if the oxide aperture of the same long-wavelength VCSEL under reference is reduced from 20 μm to 12 μm, then the incrementation in the carrier

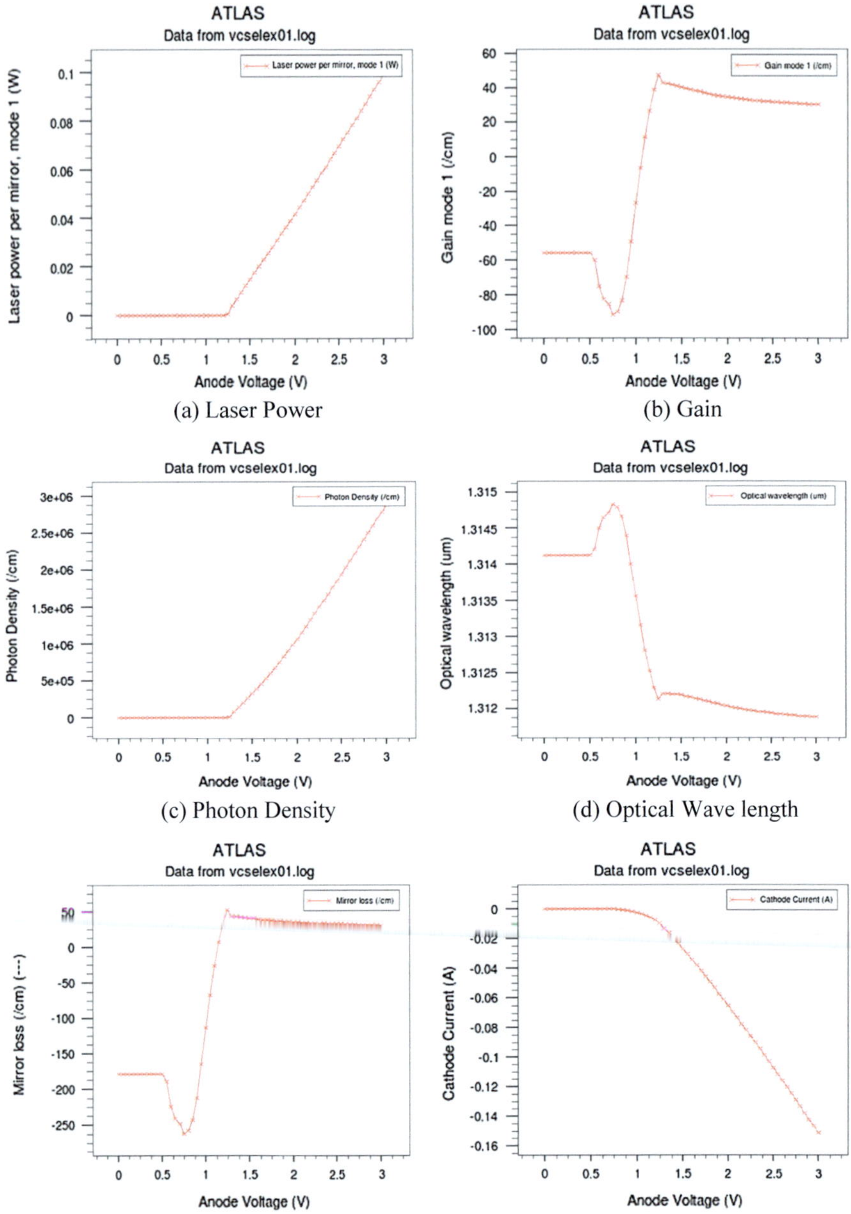

(a) Laser Power (b) Gain (c) Photon Density (d) Optical Wave length (e) Mirror Loss (f) Cathode Current

Fig. 4 Laser power, gain, photon density, optical wavelength, mirror loss, cathode current versus applied bias

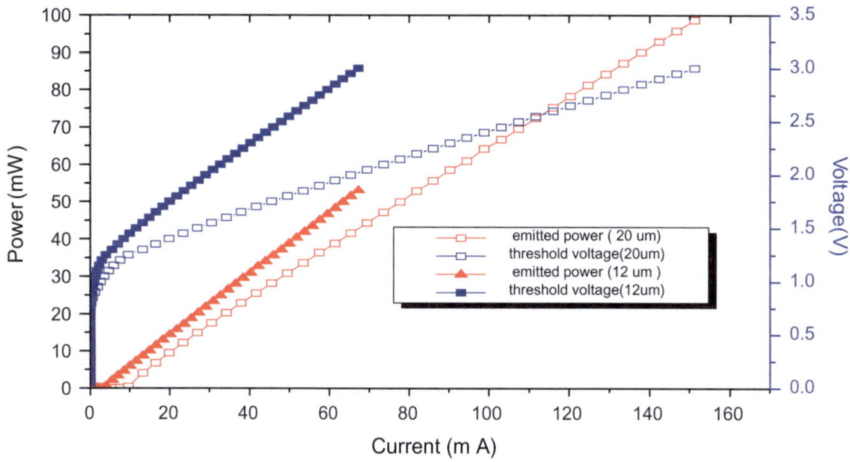

Fig. 5 Laser power, voltage versus current (LIV) characteristics of 1310 nm VCSEL

and photon density rates and subsequently reduction in the power, threshold current, and gain have been observed, as discussed in next para, in Fig. 5 [8, 54–57].

Moreover, the comparative light–current–voltage characteristics of 1310 nm emitting devices with different oxide apertures, i.e., 12 and 20 μm, is shown in Fig. 5. In this analysis, a comparison of electrical and optical characteristics of 1310-nm-long-wavelength emission of VCSEL using simulation tool is presented. The single-mode optical powers of 53 mW at 12 μm and 98.6 mW at 20 μm oxide aperture of 1310 nm emitting device have been obtained with a maximum diode voltage of 3 V. The obtained results demonstrate that by increasing the oxide aperture of the device under reference provides the better emitted or laser power.

The next section of present communication focuses on future direction of VCSELs summarized with concluding remarks.

5 Conclusion and Future Prospects

Presently, VCSEL play a essential role in every walk of life such as science and technology, research and development, consumer and industrial, medical, military/surveillance, telecommunication, etc. due to intrinsic structure of array formation, circular beam emission with small divergence, large output power, low- threshold operations, high modulation bandwidths, etc. Nowadays, the commercialization of VCSELs with above-discussed performances and numerous applications make it a strong contender over conventional edge-emitting lasers. In the present paper, the electrical and optical characteristics of 1310-nm-long-wavelength VCSEL of oxide aperture 20 and 12 μm are analyzed in industrial-specific commercial software package. The single-mode optical powers of 53 mW at 12 μm and 98.6 mW at

20 μm oxide aperture of 1310 nm emitting device have been obtained with a maximum diode voltage of 3 V and reflectivity of DBR mirrors greater than 99%. The obtained results confirm that numerical simulation presents a suitable platform to calculate and optimize diverse design parameters of VCSELs without increasing fabrication cost and time. The authors observed that if the oxide aperture of the same VCSEL is reduced from 20 to 12 μm, it obtains the increased carrier and photon density rates and subsequently reduces the threshold current and gain as shown in Fig. 5.

After deliberation on past, present of short- and long-wavelength VCSELs, it is evident that predictions are not easy to make when the things are concerned with future. The Si-based memory technology should be used for optical storage in future because from its existence, optical data storage has been listed as a future VCSEL application. High-speed and high capacity data transmission are in great demand for future optical telecommunication, computer memory systems, and mobile communications. Moreover, future paths are integration of VCSEL, PD, and actuator with MEMS technology and used for the development and manufacture of portable, cost-effective biomedical diagnostic tools.

References

1. Hall RN, Fenner GE, Kingsley JD, Soltys TJ, Carlson RO (1962) Coherent light emission from GaAs junctions. Phys Rev Lett 9:366–369
2. Iga K (2000) Surface emitting laser-its birth and generation of new optoelectronics field. IEEE J Sel Top Quantum Electron 6(6):1201–1215
3. Soda H, Iga K, Kitahara C, Suematsu Y (1979) GaInAsP/InP surface emitting injection lasers. Jap J Appl Phys 18:2329–2330
4. Chow WW, Choquette KD, Crawford MH, Lear KL, Hadley GR (1997) Design, fabrication and performance of infrared and visible vertical-cavity surface emitting lasers. IEEE J Quantum Electron 33(10):1810–1823
5. Koyama F, Kinoshita S, Iga K (1989) Room-temperature continuous wave lasing characteristics of GaAs vertical cavity surface-emitting laser. Appl Phys Lett 55:221–222
6. Jewell J, Harbison J, Scherer A, Lee Y, Florez L (1991) Vertical-Cavity surface emitting lasers: Design, growth, fabrication, characterization. IEEE J Quantum Electron 27(6):1332–1346
7. Babic DI, Streubel K, Mirin RP, Margalit NM, Bowers JE, Hu EL (1995) Room temperature continuous wave operation of 1.54 μm vertical cavity lasers. IEEE Photonics Technol Lett 7 (11):1225–1227
8. Piprek J, Babic DI, Bowers JE (1996) Numerical analysis of 1.54 um double-fused vertical-cavity lasers operating continuous-wave up to 33 °C. Appl Phys Lett 68(19):2630–2632
9. Jayaraman V, Geske JC, McDougal MH, Peters FH, Lowes TD, Char TD (1998) Uniform threshold, continuous wave, single mode 1300 nm vertical cavity lasers from 0 to 70 °C. Electron Lett 34(14):1405–1407
10. Qian Y et al (1997) Long wavelength (1.3 μm) vertical cavity surface emitting lasers with a wafer bonded mirror and an oxygen implanted confinement region. Appl Phys Lett 71(1):25–27

11. Lee Y, Jewell J, Scherer A, McCall S, Harbison J, Fiorez L (1989) Room-temperature continuous-wave vertical cavity single-quantum-well microlaser diodes. Electron Lett 25 (20):1377–1378
12. Baba T, Yogo Y, Suzuki K, Koyama F, Iga K (1993) Near room temperature continuous wave lasing characteristics of GaInAsP/InP Surface Emitting Laser. Electron Lett 29(10):913–915
13. Grabherr M, Weigl B, Riener G, Ebeling K (1996) Comparison of proton implanted and selectively oxidized vetical-cavity surface-emitting lasers. In: Conference on lasers and electron-optics, CLEO/Europe, pp 165
14. Hayashi Y, Mukaihara T, Hatori N, Ohnoki N, Matsutani A, Koyama F, Iga K (1995) Record low-threshold index guided InGaAs/GaAlAs vertical-cavity surface-emitting laser with a native oxide confinement structure. Electron Lett 31:560–562
15. Lear KL, Choquette KD, Schneider RP Jr, Kilcoyne SP, Geib KM (1995) Selectively oxidized vertical cavity surface emitting laser with 50% power conversion efficiency. Electron Lett 31:208–209
16. vander Ziel JP, Ilegems M (1975) Multilayer GaAs-Al0.3Ga0.7As dielectric quarter wave stacks grown by molecular beam epitaxy. Appl Optics 14:2627–2630
17. Geels RS et al (1991) InGaAs vertical-cavity surface-emitting lasers. IEEE J Quantum Electron 27(6):1359–1367
18. Huffaker DL, Deppe DG, Kumar K, Rogers TJ (1994) Native oxide defined ring contact for low threshold vertical cavity lasers. Appl Phys Lett 65:97–99
19. Choquette KD, Schneider RP, Lear KL, Geib KM (1994) Low threshold voltage vertical-cavity lasers fabricated by selective oxidation. Electron Lett 30(24):2043–2044
20. Lear KL, Mar A, Choquette KD, Kilcoyne SP, Schneider RP, Geib KM (1996) High frequency modulation of oxide confined vertical cavity surface emitting lasers. Electron Lett 32:457–458
21. Koyama F (2006) Recent advances of VCSEL photonics. J Light wave Technol 24:4502–4513
22. Willner AE et al (2012) Optics and photonics: key enabling technologies. Proceeding IEEE 100:1604–1643
23. Zervas MN, et al (2014) High power fiber lasers: a review. IEEE J Sel Top Quantum Electron 20(5)
24. Richardson DJ et al (2010) High power fiber lasers: current status and future prospective. J Opt Soc Amer B 27:B63–B92
25. Michalzik R, Ebeling KJ (2003) Operating principles of VCSELs. Univ of Ulm, Optoelectronics Department
26. Tell B, Lee YH, Brown Goebeler KF, Jewell JL, Leigenguth RE, Asom MT, Livescu G, Luther L, Mattera VD (1990) High-power CW vertical-cavity top surface-emitting GaAs quantum well lasers. Appl Phys Lett 57(18):1855–1857
27. Kasten AM, Tan MP, Sulkin JD, Leisher PO, Choquette KD (2008) Photonic crystal vertical cavity lasers with wavelength-independent single-mode behavior. IEEE Photon Technol Lett 20(23):2010–2012
28. Ragunathan G (2014) Design and fabrication of vertical external cavity surface-emitting lasers. Thesis for the degree of Master of Science in Electrical and Computer Engineering, University of Illinois at Urbana-Champaign
29. Larsson A, Gustavsson JS (2013) VCSELs: fundamentals, technology and applications of vertical-cavity surface-emitting lasers. Springer-Verlag, Berlin, Germany ch. 4, pp 119–144
30. Saha AK, Islam S (2009) An improved model for computing the reflectivity of a AlAs/GaAs based distributed bragg reflector and vertical cavity surface emitting laser. Optical Quantum Electron 41:873–882
31. Leonardis FD et al (2007) Improved simulation of VCSEL distributed bragg reflectors. J Comput Electron 6:289–292
32. Mitani SM, Choudhury PK, Alias MS (2007) Design and analysis of a GaAs-based 850 nm vertical cavity surface emitting laser with different doping in the reflection regions. J Russian Laser Res 28(6):610–618

33. Gronenborn S et al (2011) High-power VCSELs with a rectangular aperture. Appl Phys B Laser Opt-Springer 105:783–792
34. Seurin JF et al (2013) High power red VCSEL arrays. Proceeding of SPIE 8639:86390O-1–86390O-9
35. Alias MS et al (2009) Comprehensive uniformity analysis of GaAs-based VCSEL epiwafer by utilizing the on-wafer test capability. J Russ laser Res 30(4):368–375
36. Huffaker DL, Graham LA, Deppe DG (1996) Fabrication of high packaging density vertical cavity surface emitting laser arrays using selective oxidation. IEEE Photon Techn Lett 8:596–598
37. Sinzinger S, Jahns J (2003) Microoptics, 2nd edn. WILEY-VCH GmbH & Co., Weinheim
38. Baili G et al (2014) Ultralow noise and high-power VCSEL for high dynamic range and broadband RF/Optical links. J Light wave Technol 32(20):3489–3494
39. Iga K (2008) Vertical Cavity Surface Emitting Laser: its conceptions and evolution. Jpn J Appl Phys 47:1–10
40. Tayahi MB, Dutta NK, Hobson WS, Vakhshoori D, Lopata J, Wynn J (1997) High power InGaAs/GaAsP/lnGaP surface emitting laser. Electron Lett 33(21):1794–1795
41. Chang C-H, Chrostowski L, Chang-Hasnain CJ (2003) Injection locking of VCSELs. IEEE J Sel Top Quantum Electron 9(5):1386–1393
42. Margalit NM, Zhang SZ, Bowers JE (1997) Vertical cavity lasers for telecom applications. IEEE Communications Magazine, Newyork, pp 164–170
43. Liu JJ, Kalayjian Z, Riely B, Chang W, Simonis GJ (2003) Alyssa Apsel and Andreas Andreou, multichannel ultrathin silicon-on-sapphire optical interconnects. IEEE J. Sel Top Quantum Electron 9(2):380–386
44. Yu SF (2003) Analysis and design of vertical cavity surface emitting laser. Wiley
45. Lamy JM, Boyer Richard S, Levallois C, Paranthoen C, Folliot H, Chevalier N, Le Corre A, Loualiche S (2008) Design of an InGaAs/InP 1.55 um electrically pumped VCSEL. Opt Quant Electron 40:1193–1198
46. Grabherr M, Gerlach P, King R, Jager R (2009) Integrated photodiodes complement the VCSEL platform. Proc of SPIE 7229, pp 72290E 1–9
47. Larsson A (2011) Advances in VCSELs for communication and sensing. IEEE J Sel Top Quantum Electron 17:1552–1567
48. Birkbeck AL et al (2003) VCSEL arrays as Micromanipulators in chip-based Biosystems. Biomed Microdevices 5(1):47–54
49. Johnson K, Brenner MH, Hogan W, Dummer M (2012) Advances in red VCSEL technology. Advances in Optical Technologies, Special Issue on Recent Advances in Semiconductor Surface-Emitting Lasers, Article ID 569379
50. Challener WA et al (2009) Gage, Heat-assisted magnetic recording by a near-field transducer with efficient optical energy transfer. Nat Photon 3:220–224
51. Ukaegbu IA et al (2013) Performance analysis of vertical and horizontal transmitter array modules using short and long wavelength VCSELs for optical interconnects. IEEE Trans Components, Packaging and Manufacturing Technol 3(5):740–748
52. Seeds AJ, Williams KJ (2006) Microwave photonics. J Lightwave Technol 24(12):4628–4641
53. Chevallier C, Fressengeas N, Genty F, Jacquet J (2012) Robust design of Si/Si3N4 high contrast grating mirror for mid-infrared VCSEL application. Optical Quantum Electron 44:169–174
54. Haurylau M et al (2007) On-chip optical interconnect roadmap: challenges and critical directions. IEEE J Sel Top Quantum Electron 12(6):1699–1705
55. Menon PS, Kandiah K, Majlis BY, Shaari S (2011) Comparison of Mesa and Device Diameter Variation in Double Wafer-Fused Multi Quantum-Well, Long-Wavelength, Vertical Cavity Surface Emitting Lasers. Sains Malaysiana 40(6):631–636
56. Ivanov PS, Rorison JM (2010) Theoretical investigation of static and dynamic characteristics of vertical cavity surface emitting lasers with incorporated two-dimensional photonic crystals. Opt Quantum Electron 42:193–213

57. Rashed Ahmed Nabih Zaki, Metawe Mohamed A (2013) Operation performance characteristics of vertical cavity surface emitting lasers (VCSELs) under high thermal neutron irradiated fields. J Rus Laser Res 34(1):1–7

Photonic Crystal Based Sensor for DNA Analysis of Cancer Detection

Sandip Kumar Roy and Preeta Sharan

Abstract This paper reports a novel method for detecting cancer with a small integrated lab-on-a-chip. The existing biosensors measure the interactions between the biomolecules that are absorbed on the surfaces of biochips. In this paper, a two-dimensional Photonic Crystal (PhC) ring resonator is proposed for DNA analysis of normal and cancerous blood cells. For modeling and simulation of the sensor, Finite-Domain and Time Domain (FDTD) method is used as the backend algorithm using MEEP (MIT Electromagnetic Equation Propagation) tool. Simulation shows that, for a small change in refractive index (RI) of DNA of blood, there is a change in the spectrum. Thus, the sensor can be used as a highly sensitive sensing device for the change in DNA properties because of cancer.

Keywords Cancer · Photonic crystal · Refractive index · MEEP

1 Introduction

Among the most fatal diseases, cancer is in the third position in India. By next quarter of a century, globally reported cancer cases may get doubled. Half of the reported cases will be from developing countries. Cancer is a generic term used to classify cause of a huge set of diseases. These diseases primarily have a common symptom of uncontrolled cell proliferation which cannot be controlled by the normal cell kinetics regulation mechanism. In case of cancer, a normal cell multiplies continuously without control, leading to the development of tumors or an

S. K. Roy (✉)
AMET University, Chennai, India
e-mail: sandipr@hotmail.com

P. Sharan
The Oxford College of Engineering, VTU, Bengaluru, India
e-mail: sharanpreeta@gmail.com

A. Mishra et al. (eds.), *Silicon Photonics & High Performance Computing*,
Advances in Intelligent Systems and Computing 718,
https://doi.org/10.1007/978-981-10-7656-5_9

abnormal rise in the number of dispersed cells like the blood corpuscles. Occurrence of cancer may happen in any organ or tissue of the body. Research has not established relation of cancer with age; however, mostly it is found to occur with old age. Cancer risks increase with lifestyle and exposure to cancer-causing environment like tobacco. Infections can act as a cause to cancer but as such this is not an infectious disease. Exposure to pollutants like industrial effluents, chemical waste, therapeutic drugs, and ionizing radiation can increase the cancer risk. Lifestyle habits like junk diet, cigarette, alcohol, and exposure to industrial waste attributed to 50% cancer risks [1].

Damage to DNA is the root cause of cancer. In case of a normal cell, a damaged DNA gets either repaired or the cell ceases to exist. In case of cancer, the damaged DNA continues to make new cells even though the body may not require them. This results in unwanted tissues and manifests as either tumors or cancer [2].

When altered DNA starts evolving a different set of tissues until the cells do not become invasive, the condition is said to be a tumor. If the cells start propagating to different unaffected parts of the body thus forming a different set of tissues, the condition is said to be malignant or cancer.

According to the American Cancer Society, cancer report for the year 2013–2014, globally there are around 10 million people affected by cancer out of which 6.5 lakhs are in India. About 39,620 deaths of all age group women in US are due to breast cancer [3].

In the current work, we have presented a sensor for the DNA analysis and detection of cancer. The 2D PhC ring resonator structure is designed and simulated. The ring resonator structure is used because it provides higher accuracy results. For small changes in the optical parameters of the analyte, the design provides a significant shift in output transmitted power.

2 Theory

PhC is composed of periodic dielectric nanostructures with sandwich of materials with low and high dielectric constant in 1D, 2D, and 3D. PhC affects propagation of light inside the structure. Because of the periodicity of structure, PhC causes Photonic Band Gap (PBG) where the propagation of light is completely prohibited for certain frequency ranges (band). By means of defect engineering, the band gaps become porous for certain frequencies and light can pass through the PBG [4].

PhC possesses two types of polarization by symmetry: the transverse magnetic (TM) in which the electric and magnetic fields are orthogonal to one another, and the transverse electric field (TE) in which the electric and magnetic fields are in same phase [5].

Light propagation in PhC is governed by the master Eq. (1). This is arrived at by solving Maxwell's electromagnetic equations:

$$\nabla \times \left(\frac{1}{\varepsilon}\nabla \times H\right) = \left(\frac{\omega}{c}\right)^2 H \tag{1}$$

H Photon's magnetic field
ε Permittivity
ω Angular frequency of resonance.

As per Eq. (1), the angular frequency of propagating light (ω) is dependent on the permittivity (ε) of medium. As dielectric function ε changes, the ω changes. This is the principle behind the use of PhC structures as sensor. Methods like effective refractive index, spectroscopy, and optical imaging are there but because of small input change, output sensitivity is significantly low.

Ring resonator waveguide structures are used to increase sensitivity of the sensor. In resonator structures, output power of the waveguide alters as the refractive index of the sample is changed. The variation in resonance can be captured and thus the structure can be used for sensing applications [6].

The focus of the work is to design a 2D PhC ring resonator-based lab-on-a-chip which can be used for the DNA analysis of cancer detection. The ring resonator structure is used to increase the sensitivity and quality factor of the design. The RI of normal and cancerous DNA is considered for the change in behavior of the medium of each case. Distinct displacement in output transmitted power and frequency is obtained for normal and cancerous DNA cells [7].

3 Sensor Design

The design structure of PhC sensor is a 2D square lattice ring resonator structure. The square lattice with rods in air configuration is used. This sensor is implemented using a ring resonator defect engineering. Resonators are waveguides shaped in a ring structure. We place the analyte in the structure and a spectrum, which is unique for different cells in the sample and is obtained at the end of the waveguide. For analysis of time domain and frequency domain, we used the data of dielectric constants of DNA cell (normal and cancerous) from published data [8, 9] (Fig. 1).

MEEP/MATLAB software used to develop source code for modeling and designing of PhC waveguide. The MEEP tool uses scripting language where a code is written and simulation is carried out for band gap structures (frequency domain) and transmission spectrum with varying time period (time domain). Results obtained are represented in graph and analyzed.

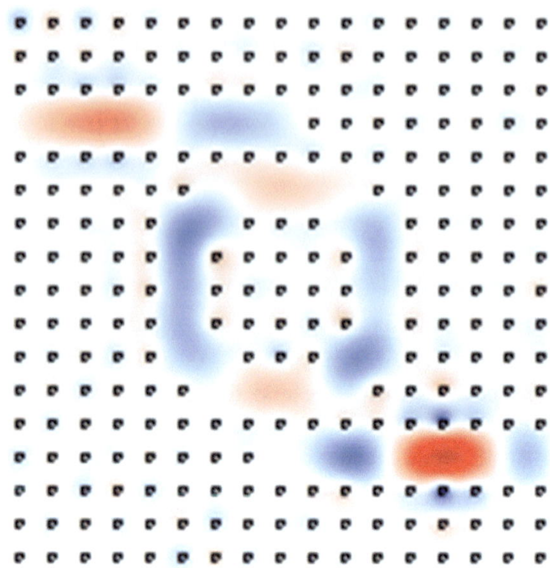

Fig. 1 2D PhC-based biosensor design using ring resonator and rods in air configuration

The design parameters considered are

1. Lattice constant (a) = 1
2. Rods in air with radius (r) = 0.2 μm
3. Silicon rods with dielectric constant = 12
4. Rods height = Infinity
5. Light's wavelength = 1555 nm
6. Dielectric constant of air is replaced with that of sample.
7. Gaussian pulsed light source used.

Detailed study has been done on variation in medium properties of DNA for normal and cancerous cells. The refractive index of each case is taken as input for MEEP.

4 Design and Modeling

Software package MEEP from MIT used for design is a free software for FDTD simulation to model electromagnetic systems, along with MPB. In this technique, discrete grid is drawn in the available space. Using discrete time steps, with time the fields are evolved. As the grids and the time steps are made further refined, a closer approximation for the continuous equations is obtained.

In MEEP, FDTD is implemented to solve time-varying Maxwell's equation. In FDTD method, a variable or constant spacing of rectangular grid points is discretised. FDTD method helps in solving Maxwell's equations directly. MPB is a frequency-domain analysis tool and implements the electromagnetic modes of periodic dielectric structures. MPB does a direct computation of the eigenstates values of Maxwell's equations. Each field computed has a definite frequency. MATLAB is used to plot graphs. In RSOFT photonic suite software, RSoft CAD is a program for passive device simulation. Band Solve module is used in our work.

5 Experiment and Results

In this paper, variation of output amplitude is compared with the introduction of normal cell and infected cell. The spectral results are simulated and tabulated for normal and cancerous DNA cells [10]. Figures 2, 3, 4, and 5 show the transmitted and reflected spectrum of normal and cancerous DNA. The result of simulation shows shift in wavelength for both normal and cancerous DNA. The peak frequency obtained for normal DNA is 275 and for cancerous DNA is 259 MEEP unit. Thus, the PhC sensor has a high-quality factor.

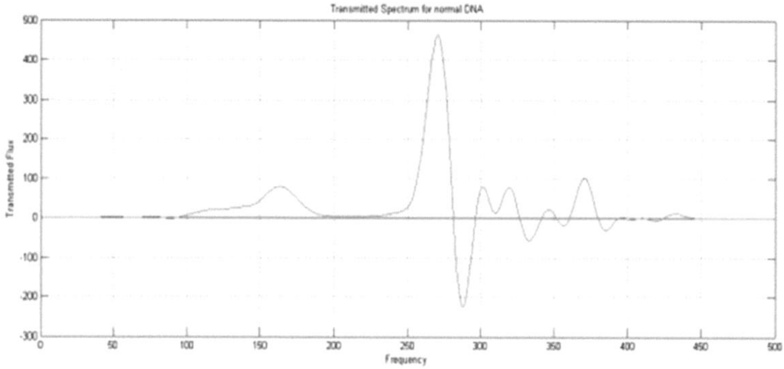

Fig. 2 Normal DNA—transmitted spectrum

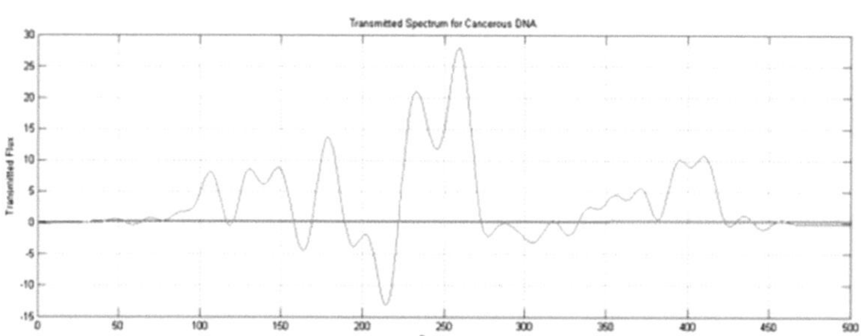

Fig. 3 Cancerous DNA—transmitted spectrum

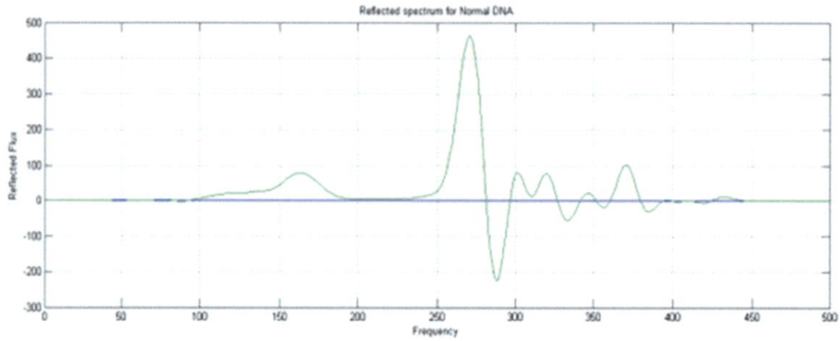

Fig. 4 Normal DNA—reflected spectrum

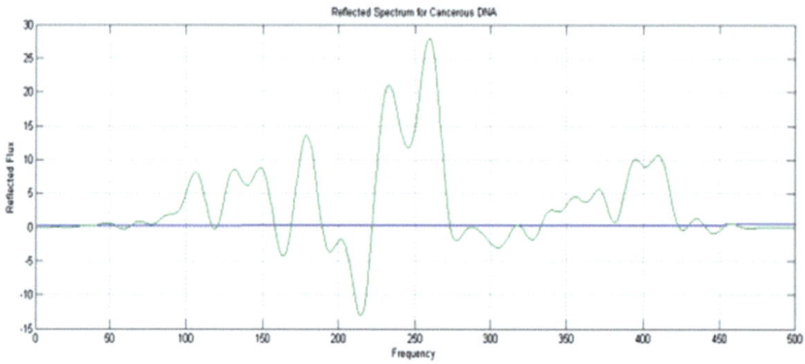

Fig. 5 Cancerous DNA—reflected spectrum

Fig. 6 Cancerous versus normal DNA—overlapping transmitted spectrum

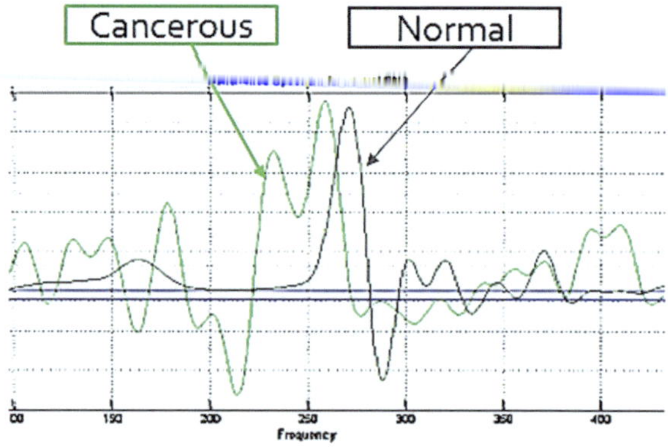

Figure 6 shows overlapping transmitted spectrum of DNA for normal versus cancerous cells. It has been shown that there was a shift in wavelength between normal and cancerous DNA. The graph in green indicates the transmitted flux value for the optical properties of the cancer-affected DNA [11, 12].

6 Conclusion

The work presented in this paper involves design of a PhC-based sensor to detect cancerous DNA. A 2D PhC ring resonator structure designed with rods in air configuration. For normal and cancerous DNA, transmission spectrum is taken using MEEP. Result shows distinct output spectrum for normal and cancerous DNA with a frequency shift of 16 MEEP unit. Thus, the work established that the distinct spectrum can be used as a signature for detection of cancerous DNA.

References

1. Steckl AJ, Spaeth H, You H, Gomez E, Grote J, DNA as an optical material
2. The National Human Genome Research Institute fact sheet Deoxyribonucleic Acid (DNA)
3. American Cancer Society (2014) Cancer Facts & Figures 2014. American Cancer Society, Atlanta, GA
4. Exploration—life through the lens www.exploratorium.edu
5. Johnson SG, Joannopoulos JD (2001) Introduction to photonic crystals. Massachusetts Institute of Technology, June 2001
6. Effects of refractive index on near-infrared tomography of the breast—PUBMED
7. Australian Research Council Centre of Excellence for Nanoscale Bio-photonics
8. Economics of Mammography in Norway, Shahid Zaman Institute of Health Management and Health Economics Medical Faculty University of Oslo
9. Jagannath RPK, Yalavarthy PK (2010) Approximation of internal refractive index variation improves image guided diffuse optical tomography of breast. IEEE Trans Biomed Eng 57 (10):2560–2563
10. The New Genetics, a publication of the National Institute of General Medical Sciences
11. The Virtual Genetics Education Centre, by the University of Leicester, offers information on DNA, genes, and chromosomes
12. Robinson S, Nakkeeran R Photonic crystal ring resonator based optical filters

Real-Time QoS Performance Analysis for Multimedia Traffic in an Optical Network

P. Piruthiviraj, Preeta Sharan and R. Nagaraj

Abstract Different multimedia data transfer analysis through the optical network with a routing table is performed by using the conversion to a next-generation IPv6-based protocol. IPv4 suffers from some drawbacks that may be preventing the growth of the Internet. IPv6 is developed to solve the several issues of the IPv4 such as delay, jitter, error, latency, dropped packets, address depletion, security, research, and extensibility. Hence, by using it will also give an expansion of the capabilities of the Internet and provide a variety of valuable conditions, including end–end mobile applications. The ideas are implemented for the transfer of multimedia data on the connection-oriented network TCP (Transmission Control Protocol). Different multimedia data used are image, audio, and video which is downloaded and streamed to the client systems of the established optical network. The file size is taken for the experimental purpose of image, audio, and video is 20 Mb respectively. Hence, we have analyzed with the Wireshark tool at the client systems with the Ubuntu 14.04 version.

Keywords IPv4 · IPv6 · TCP

1 Introduction

Real-time streaming of multimedia is popular with a high demand in Internet and there are many applications yet to meet this demand. Multimedia communication allocates a large bandwidth through the server in live streaming from the source point of access [1]. Packet loss is an important issue to be considered in data

P. Piruthiviraj (✉) · P. Sharan
The Oxford College of Engineering, Bengaluru, India
e-mail: prithivi.eie@gmail.com

P. Sharan
e-mail: sharanpreeta@gmail.com

R. Nagaraj
Dhirubhai Ambani-IICT, Gandhinagar, India

A. Mishra et al. (eds.), *Silicon Photonics & High Performance Computing*,
Advances in Intelligent Systems and Computing 718,
https://doi.org/10.1007/978-981-10-7656-5_10

transmission in case of mobile applications with the vast developments in small-sized embedded processors. Based on the infrastructure needs for a huge number of addresses which is a highest priority while using several tools as Java programming support. IPv6 plays a vital role.

IPv6 enhances the reallocation of addresses from the IP-based network (3.40282E+38 addresses) [2] to the configured networks in multiple orders with respect to IPv4 (4294967296 addresses) [3]. As such the device requires more address lines from the microprocessor or because of the further resources of the temporary memory files such as flash memory, in order to support IPv6-based network [4]. The IPv4 and IPv6 configured clients are supported by the server system configured with the IPv6 which also act as a dual-stack system and supports both. In case of IPv4 and IPv6 configured clients but IPV4 server system is able to support only the IPv4 server client and it does not support the IPv6 client. The above mentioned characteristics of IPv6 support a lot for the IPv4 network-based communication [5, 6]. By using the "Network Codes" from our routing table, even the systems configured as IPv4 connects to the communication link similar to IPv6.

The QoS for an IP-based networks in multimedia data streaming is calculated using the following parameters such as delay, speed, and time of data streaming. Then supporting of multimedia services is complex for IP-based networks and thus QoS is quite Challenging task.

2 IPv4 Communication

In the present IPv4 addressing protocol used on the Internet, limited public IPv4 addresses are used and has no more IPv4 addresses due to wider usage of address. IPv4 addresses for new customers are allocated based on the current address utilized.

3 IPv6 Communication

IPv6 was developed around 1990 from the Internet Engineering Task Force (IETF) due to limited address available in IPv4 [7]. IPv6 and IPv4 will coexist, since IPv4 to IPv6 transition is to be done. IPv4 and IPv6 vary with address size and has 128 bits [8]. IPv6 and IPv4 are used in Application Layer Gateways with translation for end-to-end support.

4 Simulation Results and Analysis

Out of the four computers, two are configured as host computers and the other two are configured as routers using routing commands on Ubuntu-14.04 platform as shown in Fig. 1. A source host and a destination host are configured. From source

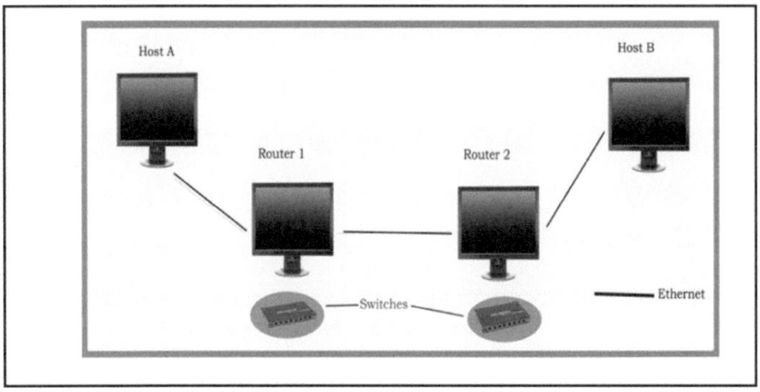

Fig. 1 Implementation diagram

hosts text, audio and bulk video data are sent and then links transfers the data with the router in Mbps. Routing tables are used for the purpose of converting the Ubuntu systems to function as routers. The optical fiber cable of different lengths results in attenuation when cable length varies. The results are verified with various optical components like couplers, attenuators, and multiplexers. Multiplexed data from Source Host 1 is transferred using a multiplexer through a bottleneck link and received at the destination host as a de-multiplexed output.

Wireshark is a tool available for Ubuntu systems, which is an open-source packet analyzer for network troubleshooting, software, and communications protocol development [9]. Wireshark analyses the structure of different networking protocols where the data can be analyzed in a network connection. Wireshark provides a GUI for the network data. Apache web server for secured authentication schemes is maintained by the server-side programming language support [10]. In Apache installation to many websites, Virtual Hosting is used.

First, the whole network is configured to be IPv4 multimedia traffic data like image, audio, and video are transmitted from server to client. The resulting analysis is as shown below with the image, audio, and video is downloaded and streamed to the client systems. So, the following figure gives the complete view of the IPv4 network from Figs. 2, 3, and 4.

Figure 2c gives the Wireshark output for the image file streamed at the client system. The graph is plotted by taking time along *X*-axis and number of packets on *Y*-axis with a unit of packet per tick. The Wireshark result analysis of the audio and video, respectively, for the client systems is shown in Fig. 3.

Second, the network was configured to be IPv6 with the same multimedia traffic data like image, audio, video are transmitted to the client systems from the server. Various multimedia analysis is shown below with the downloading stream is carried out at the client systems in the established IPv6 multimedia network. So the figures below from Figs. 5, 6, and 7 are analyzed with the time and bytes of the different data.

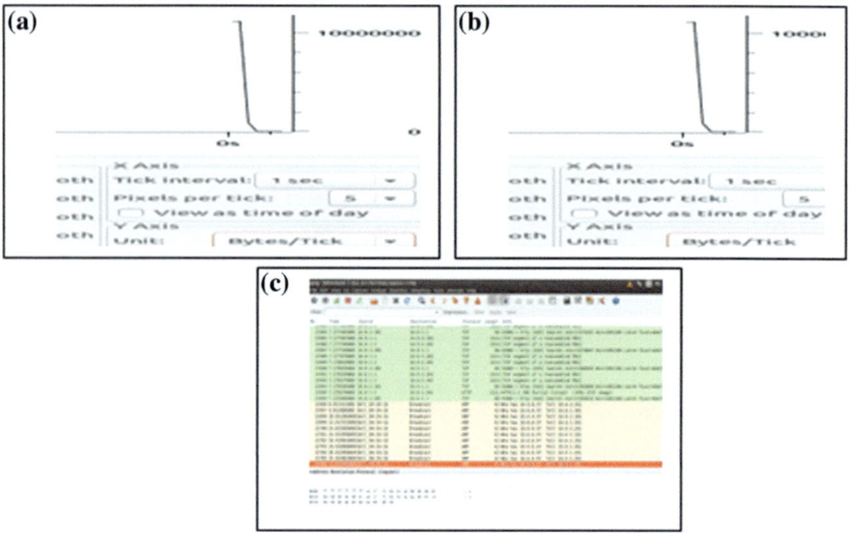

Fig. 2 IPv4 network method (image file). **a** Downloading operation. **b** Streaming operation. **c** Wireshark output

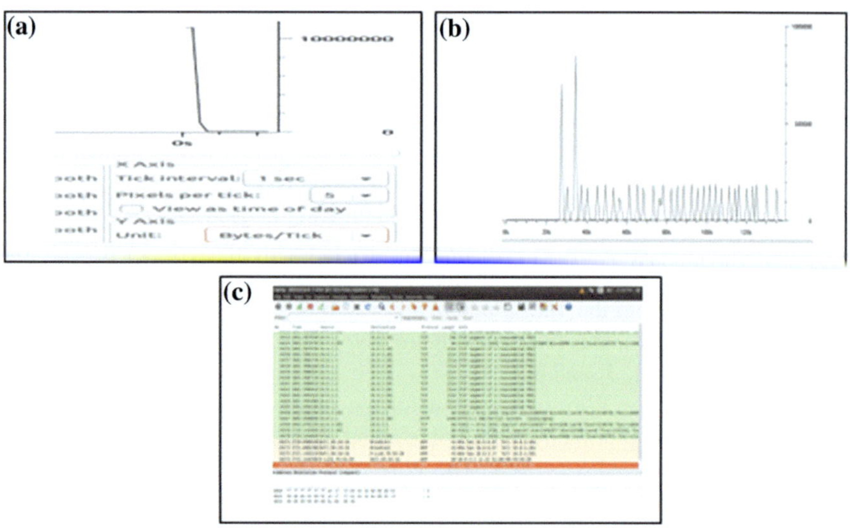

Fig. 3 IPv4 network method (audio file). **a** Downloading operation. **b** Streaming operation. **c** Wireshark output

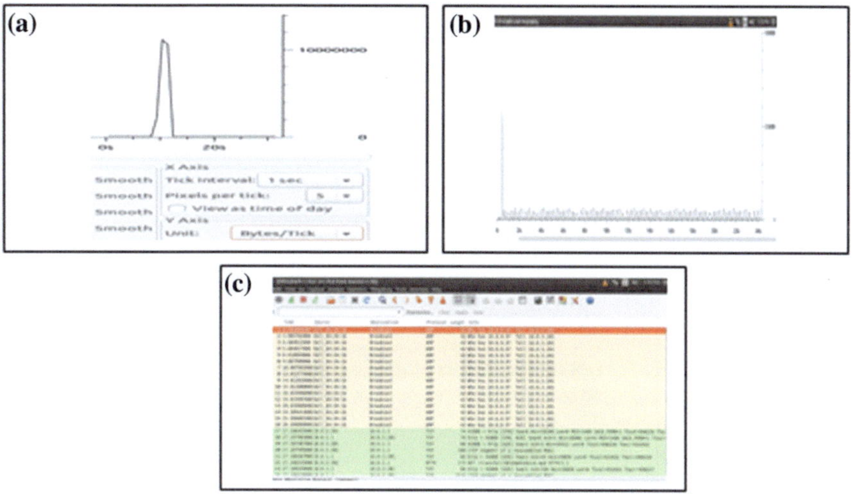

Fig. 4 IPv4 network method (video file). **a** Downloading operation. **b** Streaming operation. **c** Wireshark output

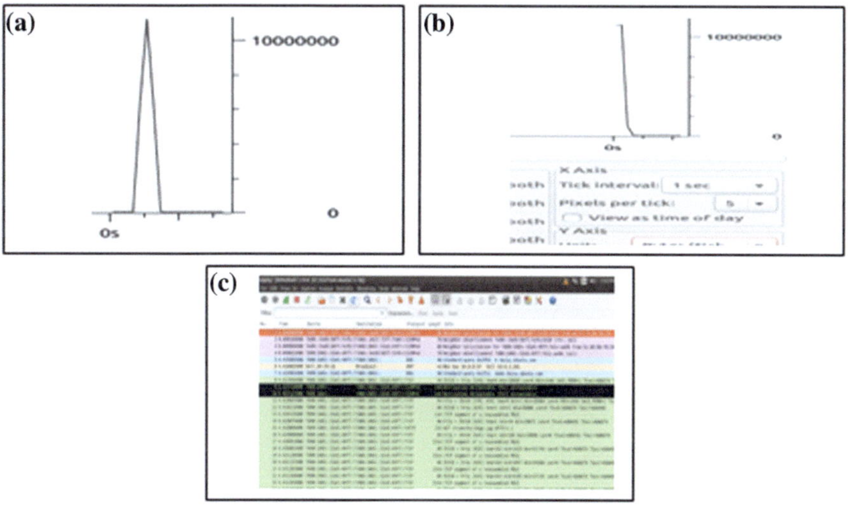

Fig. 5 IPv6 network method (image file). **a** Downloading operation. **b** Streaming operation. **c** Wireshark output

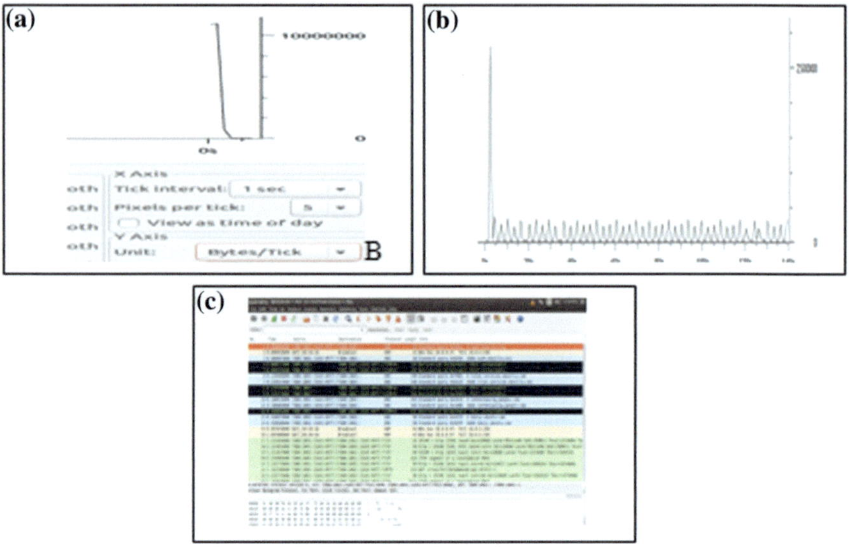

Fig. 6 IPv6 network method (audio file). **a** Downloading operation. **b** Streaming operation. **c** Wireshark output

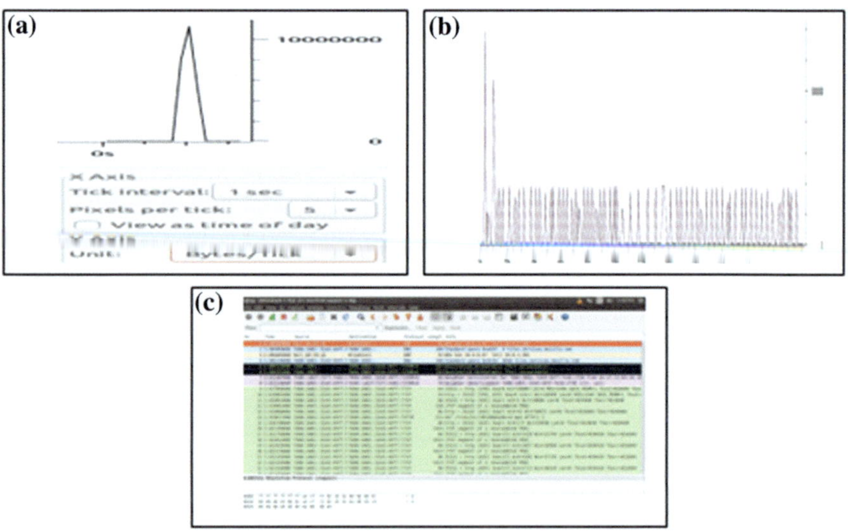

Fig. 7 IPv6 network method (video file). **a** Downloading operation. **b** Streaming operation. **c** Wireshark output

5 Conclusion

The systems are configured as routers for the analysis of IPv4–IPv6 transitions in the network to transfer multimedia data. Using the Apache server, for the various multimedia data analysis such as for image, audio, and video indicates about the downloading and streaming time in reference to the number of bytes. The multimedia data streaming is calculated as variation in the number of bytes during the different transfer of data and the downloading time. The values of the time at which it is streamed varies with image, audio, and video are 2, 3, and 12 s, respectively, for IPv4 and 2, 3, and 11 s for IPv6 respectively. The streaming of the same image, audio, and video of 20 Mb each has the value of 9, 140, and 240 s for IPv4 network and 7, 120, and 220 s for IPv6 network using Wireshark network analyzer.

Due to the nonavailability of IPv4 addresses, the system address transition to IPv6 helps not to replace the older systems. Hence, further work has been estimated to implement a tunneled IPv4–IPv6 network and a mixed network and also to obtain the results using optical cables. So the analysis of these results will help in getting a clear picture of various multimedia data over the IPv4–IPv6 communication network.

References

1. Tzeng S-S, N. C. U. C. T. Dept. of Comput. Sci. & Inf. Eng., Kuo H-C, Li L-S, Yang Y-Y (2010) An efficient multicast scheme in the nested NEMO. In: Computer communication control and automation (3CA), Tainan
2. Govil J, Govil J, Kaur N, Kaur H (2008) An examination of IPv4 and IPv6 networks: constraints and various transition mechanisms. In: Southeastcon, Huntsville
3. Winter R (2008) IPv6: why move to the next generation of the internet protocol? 8 May 2008. (Online). Available: www.dell.com/innovation
4. Yaroslavtsev AF, Lee T-J, Chung YM, Choo H (2004) Performance analysis of IP-based multimedia communication networks to support video traffic. In: Computational science—ICCS 2004. Novosibirsk, Springer, Berlin, pp 573–576
5. Oh H, Kijoon C, Hyochan B, Na J (2006) Comparisons analysis of security vulnerabilities for security enforcement in IPv4/IPv6. In: Advanced communication technology ICACT, Phoenix Park
6. www.ipv6.com, "Ipv6.com Inc," Ipv6.com Inc, 6 June 2012. (Online). Available: www.ipv6.com (2015)
7. Tadayoni R, Henten AH (2012) Transition from IPv4 to IPv6. In: 23rd European regional ITS conference, Vienna, Austria, 2012
8. Dutta C, Singh R (2012) Sustainable IPv4 to IPv6 transition. 2(10):298–305
9. Wireshark (2015) Chapter 1. Introduction. Wireshark, 20 Jan 2004. (Online). Available: www.wireshark.org
10. Lencse G, Repas S (2013) Performance analysis and comparison of different DNS64 implementations for Linux, Open BSD and FreeBSD. In: AINA '13 proceedings of the 2013 IEEE 27th international conference on advanced information networking and applications, Washington

An Optimized Design of Complex Multiply-Accumulate (MAC) Unit in Quantum Dot Cellular Automata (QCA)

G. Ambika, G. M. Shanthala, Preeta Sharan and Srinivas Talabattula

Abstract Multiply-accumulate (MAC) unit finds the large number of applications including computers and processors. In this chapter, we are implementing the optimized design of MAC unit using emerging nanotechnology called Quantum dot Cellular Automata (QCA). The multiplexer is the key unit in the design of MAC and we have achieved 34.78% reduction in total area in the design of multiplexer using QCA. As QCA has advantages such as reduced area and cell count, simulation time, less complexity with low power consumption, quality output with high efficiency is achieved.

Keywords QCA · Multiply-accumulate unit · Multiplexer

1 Introduction

CMOS technology played a vital role in the semiconductor industry in the past few decades by implementing very large-scale integrated devices. Due to the physical limits of this technology, researchers have switched to novel nanotechnology such as Quantum dot Cellular Automata (QCA), which is a top emerging technology

G. Ambika (✉) · P. Sharan
Department of ECE, The Oxford College of Engineering, Bangalore, India
e-mail: ambikagumpe1@gmail.com

P. Sharan
e-mail: sharanpreeta@gmail.com

G. M. Shanthala
Department of ECE, KSSEM, Bangalore, India
e-mail: Shanthala.g.m@kssem.edu.in

S. Talabattula
Applied Photonics Lab, IISC, Bangalore, India
e-mail: tsrinuiisc@gmail.com

A. Mishra et al. (eds.), *Silicon Photonics & High Performance Computing*,
Advances in Intelligent Systems and Computing 718,
https://doi.org/10.1007/978-981-10-7656-5_11

with potential application in future computers. QCA is based on Coulombic interaction instead of the current used in CMOS with advantages such as high switching speed, extremely low power consumption without using transistors. QCA cell is the building block of QCA devices, which consists of four quantum dots with two mobile electrons.

The basic building block of QCA circuit is majority gate and using majority gate and inverter basic gates can be built. Complex logic circuits are constructed very easily with the help of QCA basic gates. In this chapter, we are implementing the design of 2:1 multiplexer using QCA which is the basic block of MAC. In real-time DSP systems, many applications such as speech codecs, MP3, etc., require MAC operations. In this chapter, we propose a new design methodology for 2:1 multiplexer using QCA Designer software with a decreased number of cells giving efficient output.

2 Quantum Dot Cellular Automata

The square nanostructure of electron wells which confines free electrons is basically called as QCA. Each cell constitutes four quantum dots, which are situated at the corners of a square coupled by means of tunneling barrier. As physical occupancy of quantum dots varies in nanometer range, they are capable of entrapment of electrons in three dimensions. Two electrons occupy nonadjacent corners representing two polarizations by Coulombic repulsion. Here, binary "Low" is represented by $P = -1$ and $P = +1$ represents binary "High". Figure 1a, b represents how charge designation of $P2$ is affected by its neighbor $P1$.

Majority gate is the basic unit of QCA which consists of five cells, three inputs, one output, and a middle cell. Middle cell is named as device cell as it determines

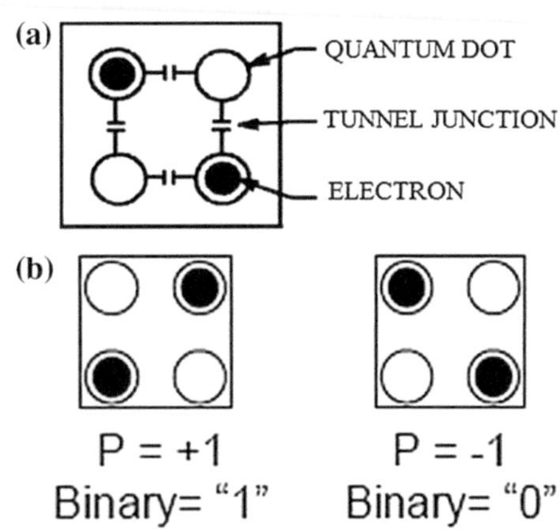

Fig. 1 **a** Basic quantum dot. **b** Polarization states

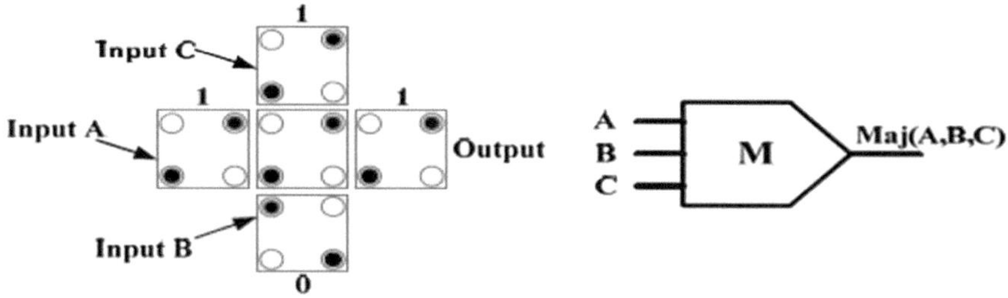

Fig. 2 Majority gate

the stable output by switching the device to major polarization. Using majority gate basic AND and OR gates can be designed by fixing one of the inputs to $P = -1$ or $P = +1$ respectively. Boolean expression for majority gate is represented as (Fig. 2):

$$\mathrm{M}\,(\mathrm{A},\,\mathrm{B},\,\mathrm{C}) = \mathrm{AB} + \mathrm{BC} + \mathrm{AC} \tag{1}$$

3 Design

Multiply-accumulate unit architecture is as shown in Fig. 3, which mainly consists of several 2:1 multiplexers. We are implementing 2:1 multiplexer which uses less area with high switching speed and low power consumption, which will lead to efficient MAC design.

The basic functional block of 2:1 multiplexer is as follows (Fig. 4).

Output of the multiplexer is based on select input. 2:1 multiplexer consists of two inputs A and B, depending upon the select line either A or B is produced at the output. Truth table is represented in Table 1, using this output expression can be given as

$$\mathrm{C} = \mathrm{A.SEL}' + \mathrm{B.SEL} \tag{2}$$

Majority gate implementation of 2:1 multiplexer is represented as (Fig. 5).

Using QCA Designer 2.0.3 multiplexer design is implemented which is the basic block for MAC unit. This design is very efficient compared to previous designs in terms of number of cells and area usage (Fig. 6).

The proposed design is compared with the previous designs and results are tabulated and plotted as (Table 2; Fig. 7).

This design is compared with the classical logic gate design and it is summarized as follows (Table 3; Fig. 8).

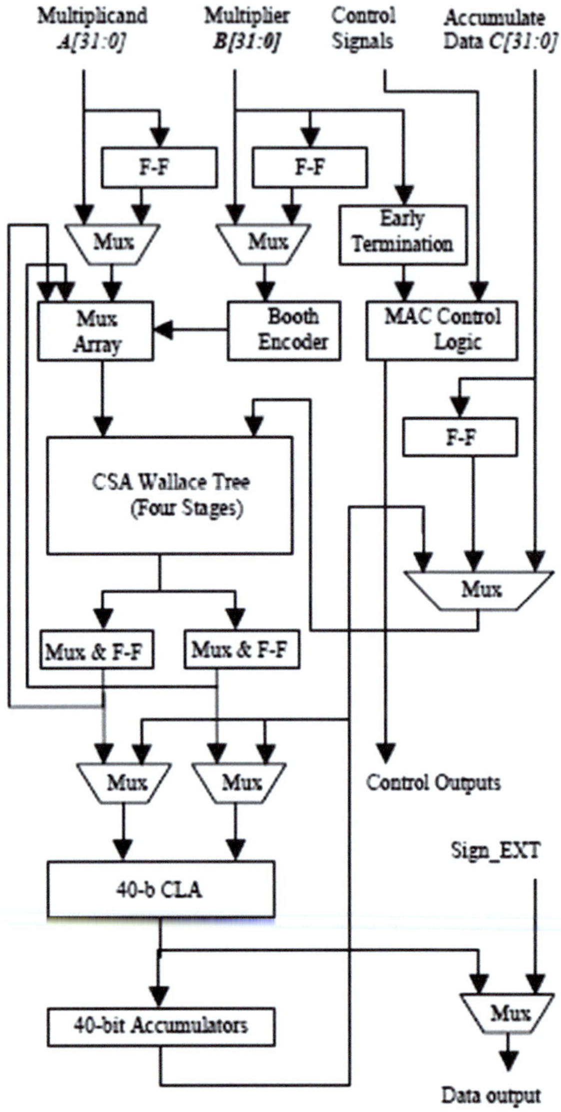

Fig. 3 MAC architecture

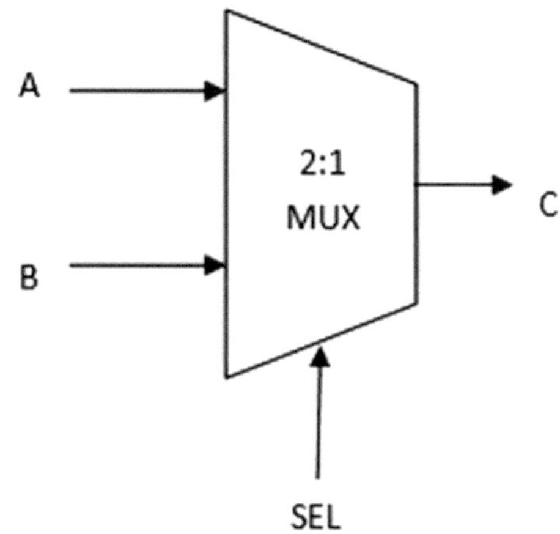

Fig. 4 Functional block of 2:1 multiplexer

Table 1 Truth table of 2:1 multiplexer

Select I/P = SEL	O/P = C
0	A
1	B

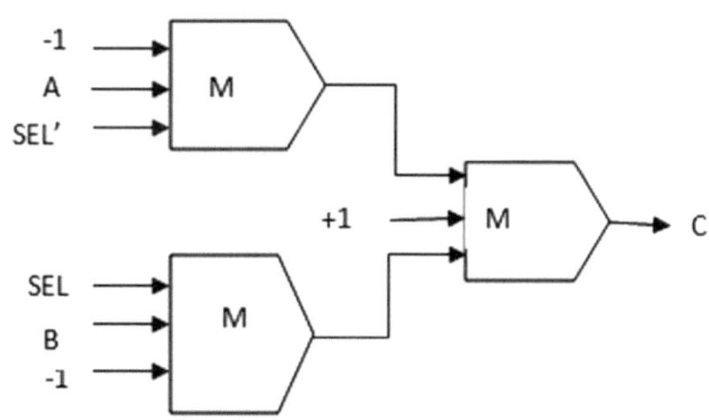

Fig. 5 Outline of the proposed 2:1 multiplexer

4 Simulation Results

Circuit functionality is verified using QCA Designer 2.0.3 [1]. Following parameters are used for bistable approximation:

- Number of samples = 12,800
- Convergence tolerance = 0.00100

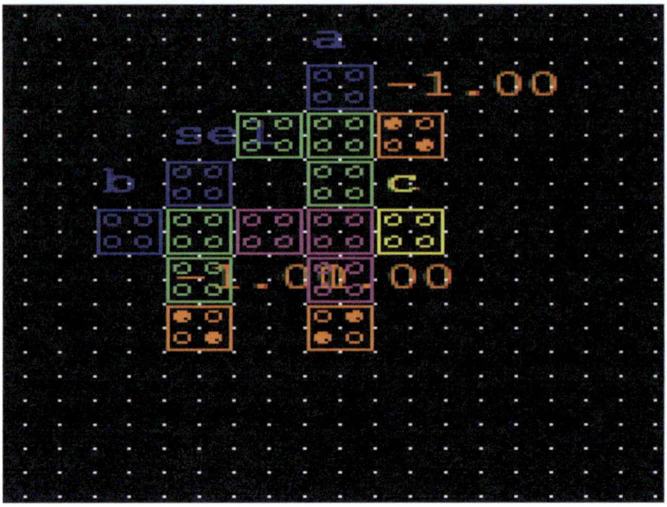

Fig. 6 Layout of 2:1 multiplexer in QCAD

Table 2 Comparison of the proposed design with some previous designs

Parameters	2:1 mux as in paper [2]	2:1 mux as in paper [3]	2:1 mux as in paper [4]	2:1 mux as in paper [5]	2:1 mux as in paper [6]	Proposed design
No. of cells	63	49	31	27	23	15
Area of cells (sq nm)	28,512	14,904	37,908	23,328	18,144	13,688
Total area (sq µm)	0.14	0.08	10,044	8748	7452	4860
Input to output delay	Seven clock zone	Four clock zone	Five clock zone	Four clock zone	Four clock zone	Four clock zone

- Radius of effect = 65 nm
- Cell size = 35 nm
- Relative permittivity = 12.9
- Clock high = 9.8e−22
- Clock low = 3.8e−23
- Clock amplitude factor = 2 (Fig. 9).

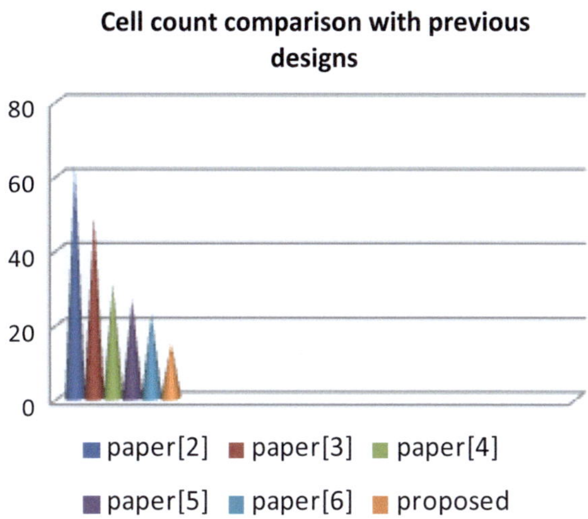

Fig. 7 Cell count comparison graph

Table 3 Comparison of cell area in the design of multiplexer using logic gates and QCAD

	Logic gates	QCAD
Cell area (sq μm)	100	0.02

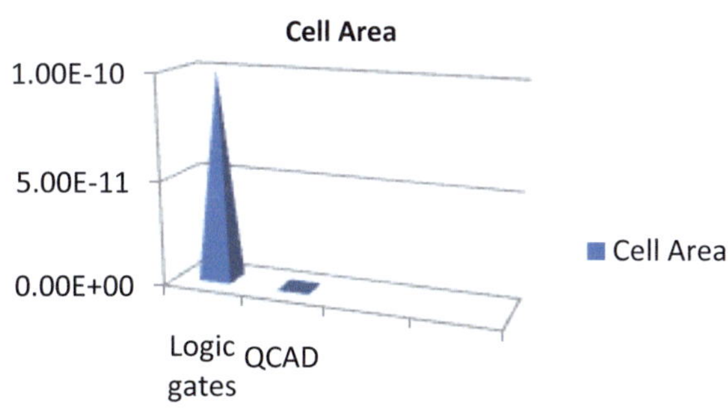

Fig. 8 Cell area comparison of QCAD result with logic gates

Simulation results shown above give the working of 2:1 multiplexer and this design is used for MAC architecture.

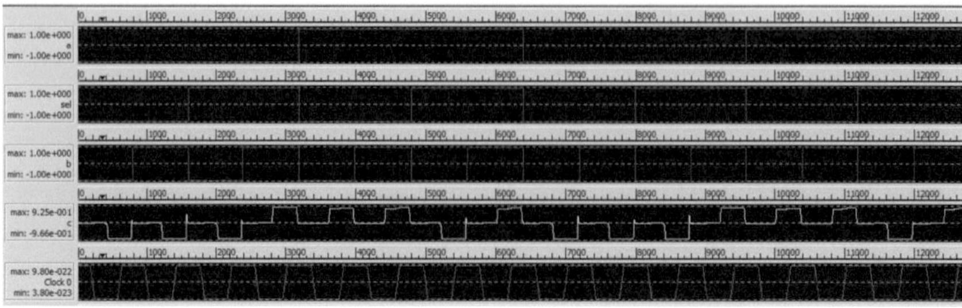

Fig. 9 Simulation results

5 Conclusion

Our proposed design of multiplexer is very simple and efficient with a remarkable reduction in the number of cells and is utilized in the architecture of MAC unit which has a large number of applications including future computers. As the design is implemented in QCA, we achieve advantages such as reduced cell area by 34.78% and it uses only 15 cells with high switching speed and low power consumption giving more efficient output when compared to previous designs. Hence, our proposed design is the promising step toward the goal of low power design.

References

1. Orlov AO, Amlani I, Bernstein GH, Lent CS, Snider GL (1997) Realization of a functional cell for quantum dot cellular automata. Science 277:928
2. Vankamamidi V, Ottavi M, Lombardi F (2008) Computer-aided design of integrated circuits and systems. IEEE Trans 27:34–44
3. Kaye P, Laamme R, Mosca M (2007) An introduction to quantum computing. Oxford University Press, Oxford
4. Mukhopadhyay D, Dinda S, Dutta P (2011) Int J Comput Appl 25:21–24
5. Ditti, Mahata S, Mitra K, Sikdar P (2009) 4th international conference on nanotechnology IEEE explore, pp 293–298
6. Hashemi S, MostafaRahimiAzghadi, Zakerolhosseini A (2012) International symposium on telecommunications, IEEE explore, 692

Modification of L2 Learning Switch Code for Firewall Functionality in POX Controller

Chaitra N. Shivayogimath and N. V. Uma Reddy

Abstract Software-Defined Network is a new evolving networking technology, separating the control plane from the forwarding plane. The Routing functionality and the API for the applications plane are on the Controller. One such application on POX controller is firewall. A modification of the Learning layer 2 switch code for POX controller is done for a tree topology of depth 3 by using mininet network emulator and the packet flow between the hosts is controlled according to the rules inserted in the Learning switch.

Keywords POX · SDN · Controller · Learning switch · Firewall

1 Introduction

Software-Defined Networking (SDN) is the new trend in the networking field. The separation of the control plane from the forwarding plane has enabled the complete programmability of the network, since the control plane and the forwarding plane are decoupled [1]. All the routing decisions are undertaken by the component of the control plane, the Controller. The forwarding function is performed by a dumb device like the Switch.

The physical test beds are very costly to implement for research and for study purposes of SDN, hence there is a need for virtualization. One such virtualization technique for SDN is the network emulation software, Mininet. Mininet allows the launching of a virtual network with switches, hosts, and an SDN controller using the limited resources on a laptop all with a single command. Mininet has all the

C. N. Shivayogimath (✉) · N. V. Uma Reddy
Department of Electronics and Communication, AMC Engineering College, Bangalore, India
e-mail: chaishivyogi@gmail.com

N. V. Uma Reddy
e-mail: nvumareddy@gmail.com

A. Mishra et al. (eds.), *Silicon Photonics & High Performance Computing*,
Advances in Intelligent Systems and Computing 718,
https://doi.org/10.1007/978-981-10-7656-5_12

three components of SDN, namely, Controller, Openflow protocol (Southbound API), and the Controller Applications (Northbound API). The architecture of SDN is shown in Fig. 1.

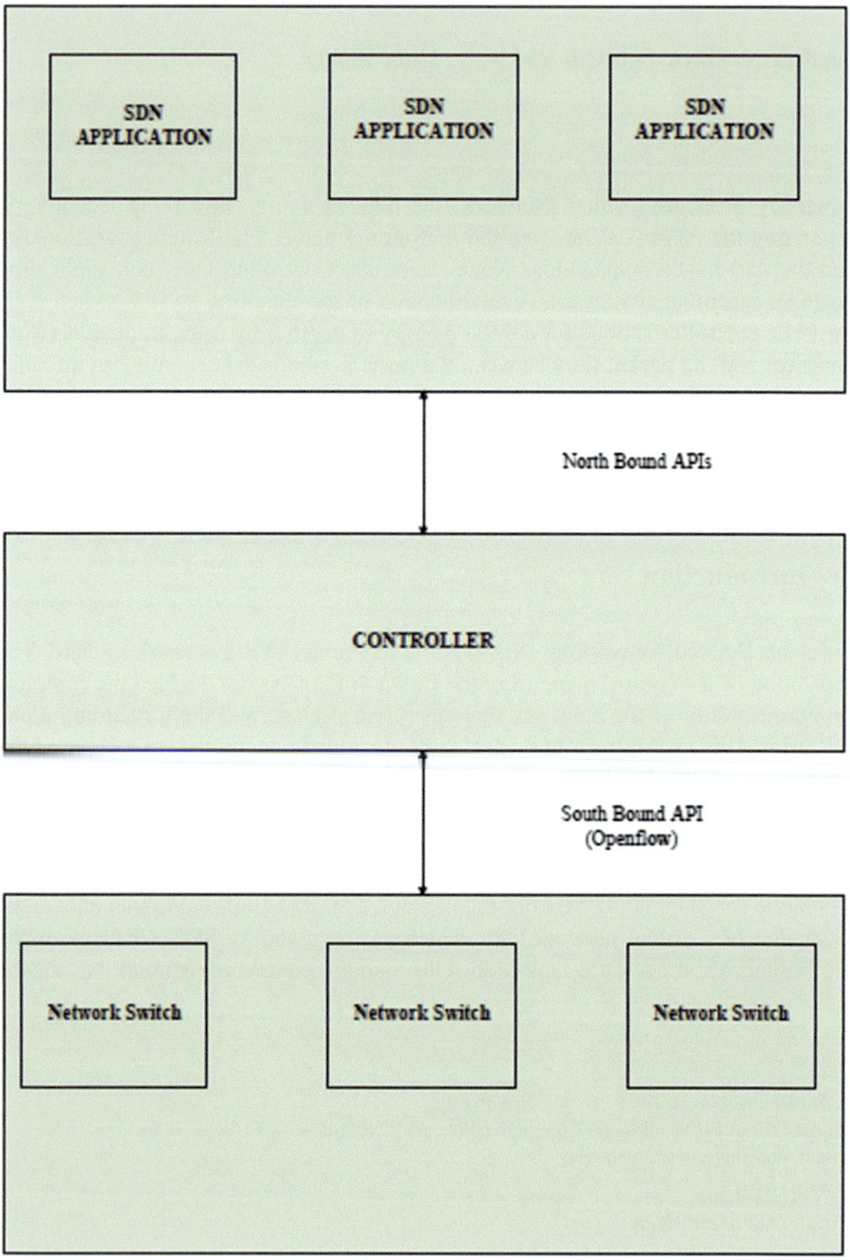

Fig. 1 SDN architecture

The most popular controller implementation is POX, which has been developed by NICIRA. POX core and its components are written in Python. Mininet also uses the POX controller. The earlier version of POX was NOX, which was written in Python and C++. Many such controllers have been researched and developed by various groups of developers, such as MUL (Kulcloud), Maestro (Rice University), Trema (NEC), Beacon (Stanford), Flowvisor (NICIRA/Stanford), and RYU (NIT, OSRG Group) [2]. The controller is chosen on the basis of programming knowledge and suitability of the controller for applications.

The forwarding action is done by the switch, which is specifically an OpenVswitch that supports openflow protocol to communicate with the controller and hence the name "Openflow Controller". The OpenVswitch is a production-quality, multilayer virtual switch. OpenVswitch can act as a soft switch running on a hypervisor or on a physical device, which can be installed on a wide range of platforms [3]. The default TCP port used by the OpenVswitch is 6633. The following three types of messages are supported by openflow [4, 6]:

(1) *Controller-to-Switch messages*: These messages are sent from the controller to the switch. The switch may or may not respond to these messages.

- Specify, modify, or delete flow entries.
- Request information about switch capabilities.
- Retrieve information like counters from the switch.
- Sending the packet back to the switch once the flow is created.

(2) *Asynchronous messages*: These messages are sent by the switch.

- Send the controller a packet that does not match the existing flow.
- Send a message to the controller about the removal of a flow due to the buffer time expiry.
- Inform the controller about the change in port or an error in switch.

(3) *Synchronous messages*: These messages are bidirectional to the switch and the controller.

- Hello messages between the controller and the switch.
- Echo messages are used to determine the latency of the controller to switch the connection and to verify that the controller to switch connection is still up.

In this chapter, the Learning switch code of the POX controller is modified to check for the source MAC address in the packet by inserting rules in the switches and allow only specific source MAC addresses for a tree topology of depth 3, thus providing a firewall kind of functionality to the controller.

2 Network Topology Creation in Mininet

The topology created for the work is a tree topology with depth 3, as shown in Fig. 2. Mininet network emulator is used to create this topology [5].

The controller is named as c0. The openflow switches are named as s1, s2, s3, s4, s5, s6, and s7, and the terminal virtual hosts are named as h1, h2, h3, h4, h5, h6, h7, and h8. Mininet VM is started in Oracle VM Virtual Box. The virtual topology shown in Fig. 2 is a tree topology with depth 3 with only one controller c0. All devices are connected via virtual Ethernet links, as labeled in Fig. 2.

The POX controller was started on a remote windows machine hosted on the network. The mininet VM was started in an Oracle VM virtual box. The mininet command used to create the tree topology is "mininet@mininet-vm: ~ $ sudo mn –topo=tree,3 –mac –controller=remote,ip=192.168.3.32,port=6633." The –mac option in the above command sets the virtual host MAC and IP addresses to small, unique, and easy-to-read IDs. The controller is set to remote IP address on 192.168.3.32 and the TCP port used is 6633. The port need not be mentioned because it takes the default of 6633. The port needs to be mentioned only if a port other than the default has to be used.

The topology creation in mininet was studied and implemented as shown in Fig. 3.

The net command output is also shown along with the topology for the link and port information in Fig. 4.

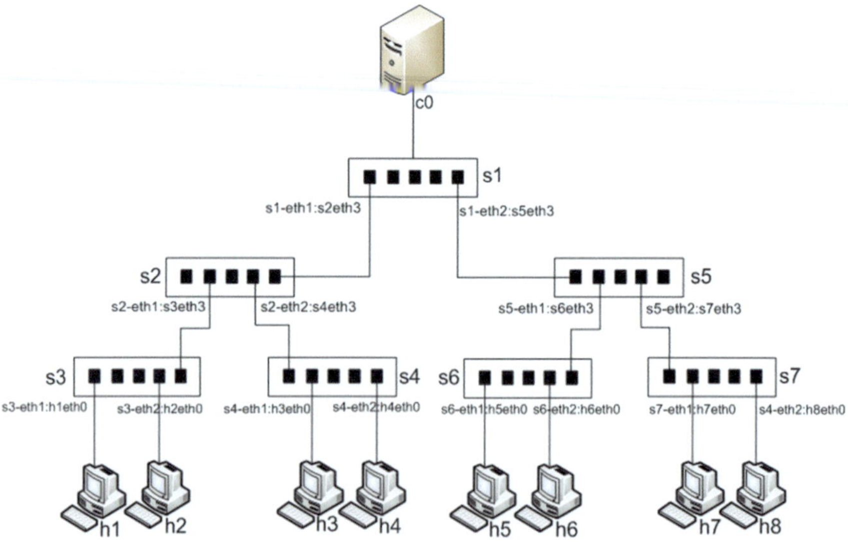

Fig. 2 A tree topology created in Mininet

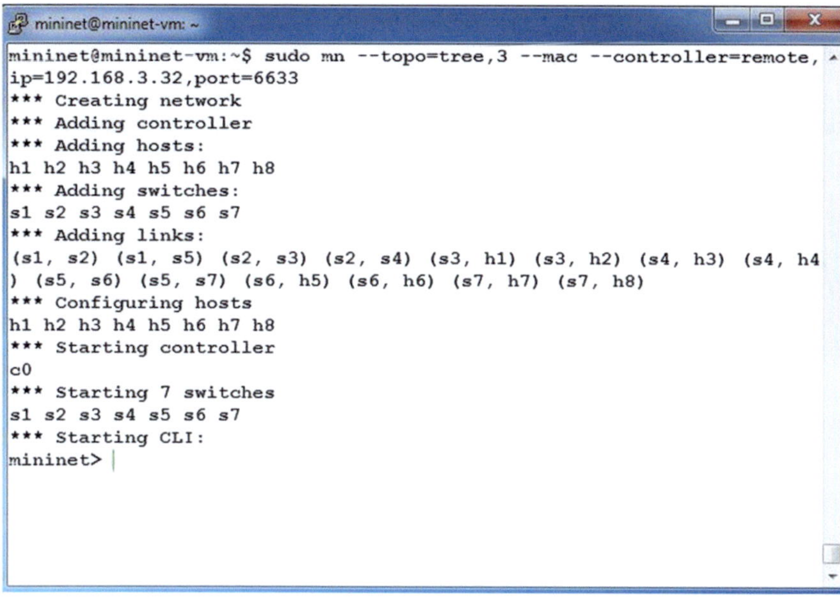

Fig. 3 Creation of Mininet topology through CLI

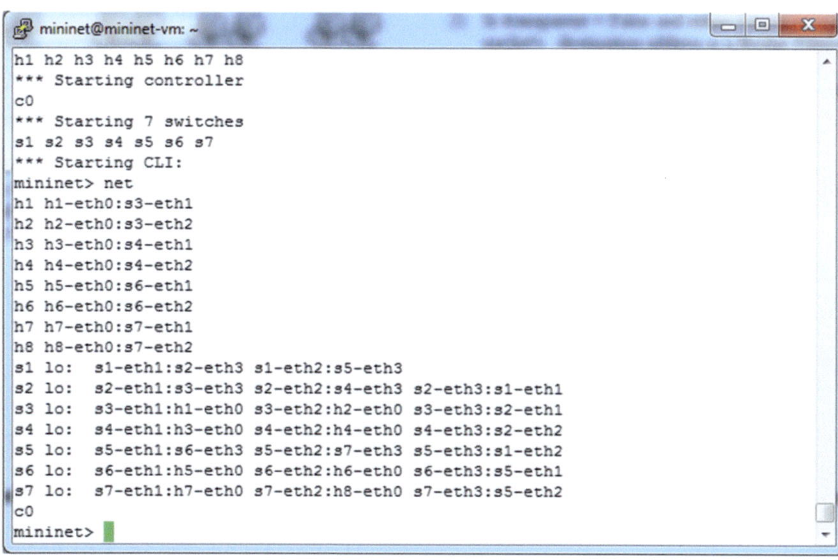

Fig. 4 The net links between the switches and the hosts are obtained by issuing net command in mininet prompt

3 Modification of L2 Learning Switch Algorithm

The L2 learning switch code is modified to perform a firewall functionality. This is done for a tree topology. The preceding subsections explain the complete flow of the modification of the code and its verification.

3.1 Learning Switch Algorithm Present in POX Controller

The learning switch algorithm present in the POX controller is given below:

(1) Source addresses and switch ports are used to update address/port table.

(2) It is checked if the packet is a transparent packet or not and check if its Ethertype is LLDP and also check if the destination address of the packet is a bridge-filtered address or not.

- If any of the above is true, then:
- The packet should be dropped and it is not supposed to forward the link-local traffic (i.e., LLDP, 802.1x)
- Else, "DONE". Go to next step.

(3) It is checked if the destination is a multicast address:

- If Yes, then:
- Flood the packet

(4) A check is done if the port for the destination address is in the address/port table or not.

- If not, then:
- Flood the packet
- Else, "DONE". Go to next step.

(5) It is checked if the output port is the same as the input port?

- If, Yes then:
- Drop packet and similar ones for a while

(6) A flow table entry is installed in the switch so that this flow goes out the appropriate port.

- Send the packet out appropriate port.

3.2 Firewall Algorithm (Based on MAC Addresses Only)

This algorithm for learning switch is modified to perform the action of a firewall, allowing only particular MAC addresses to communicate over the network through all the openflow switches in the network.

The algorithm for the firewall functionality is given below:

(1) A hash table is required to store the key:value pair of (switch, sourceMAC).
(2) Check the source MAC address against the Rule added.
(3) The table maps the (switch, source MAC) to True or False
(4) The Controller decides to drop the traffic in either of the following two conditions:
   - If there is a firewall rule matching "False".
   - If there is no firewall rule entry.
(5) The packet is forwarded if the rule matches "True".

3.3 Modification of Learning Switch Code of POX Controller to Perform the Firewall Functionality

The POX controller consists of 3 parts:

(1) Listener
(2) Control Logic
(3) Messenger

The Listener has "Packet in" and "Connection Up" messages [7]. The logic in the "Packet in" module is modified to check for the added firewall rules and then perform the usual learning switch functionality if the rule matches "True" (i.e., sending a message to switch to add a new rule in the Openflow Flow Table), else the packet is dropped.

The existing Learning Switch algorithm is modified at step 2 to check the rule.

4 Verification of Modified Algorithm and Code

First the rule was added in s3 and s7 only using the dpid of s3 and s7 [8]. This failed since the rules were not present in the intermediate Openflow switches.

In Fig. 5, a debug message is displayed for the rule found in s3, which shows that the packet was forwarded. Further, since a rule for the next switch (i.e., s2) was not found to reach the controller, the packet was found to be dropped.

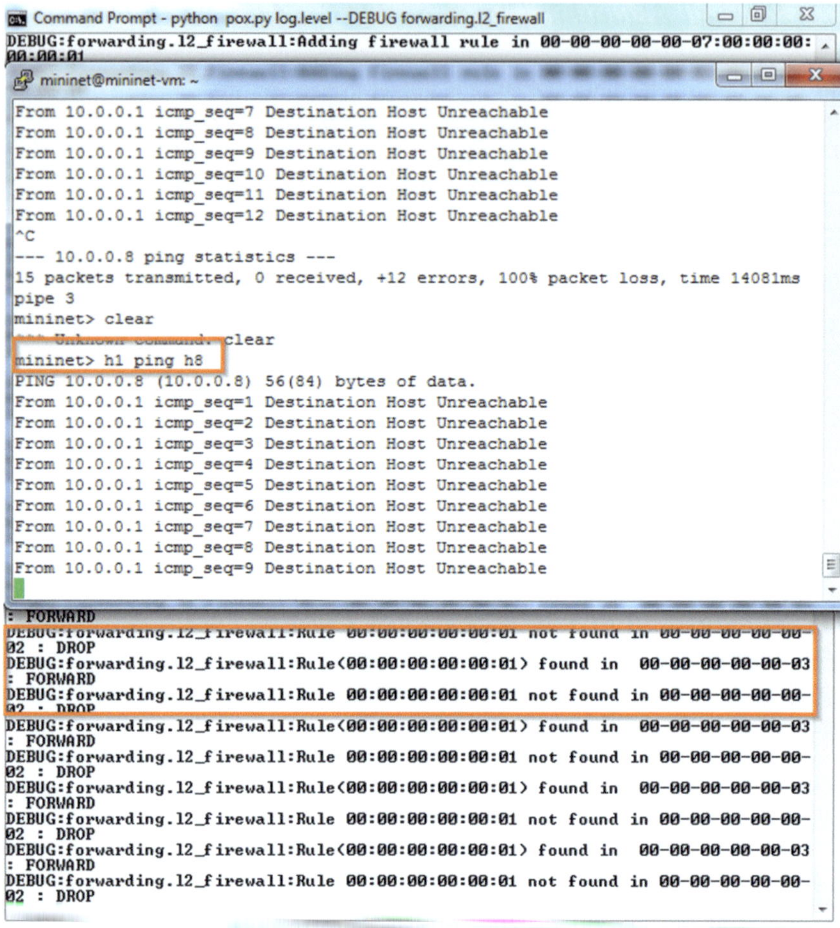

Fig. 5 Debug messages in POX controller when the rule is not found in a switch

This problem was overcome by adding the rules in all the 7 switches. When a pingall test was conducted after all modifications, only h1 and h8 were reachable, as seen in Fig. 7. The debug messages for the pingall test are shown in Fig. 6.

The rules were modified to allow communication between various other hosts, for example, communication between h2–h3–h4 and h4–h6. The POX controller when started with log level set to DEBUG, the first component of the POX controller, i.e., Listener (ConnectionUp), is executed and once the mininet topology is created, the switches get connected to the POX controller. The Info message “Connection {switch dpid} is connected” is displayed. The rules get added on all the switches, as shown in Fig. 8.

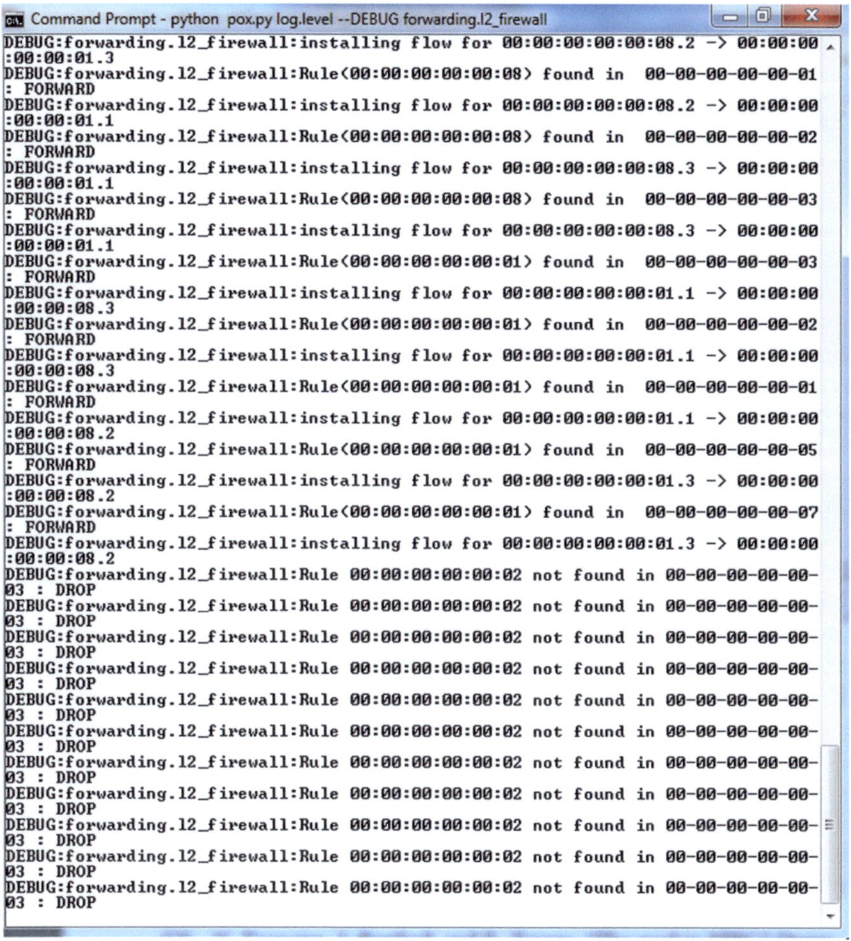

```
Command Prompt - python pox.py log.level --DEBUG forwarding.l2_firewall
DEBUG:forwarding.l2_firewall:installing flow for 00:00:00:00:00:08.2 -> 00:00:00
:00:00:01.3
DEBUG:forwarding.l2_firewall:Rule(00:00:00:00:00:08) found in  00-00-00-00-00-01
: FORWARD
DEBUG:forwarding.l2_firewall:installing flow for 00:00:00:00:00:08.2 -> 00:00:00
:00:00:01.1
DEBUG:forwarding.l2_firewall:Rule(00:00:00:00:00:08) found in  00-00-00-00-00-02
: FORWARD
DEBUG:forwarding.l2_firewall:installing flow for 00:00:00:00:00:08.3 -> 00:00:00
:00:00:01.1
DEBUG:forwarding.l2_firewall:Rule(00:00:00:00:00:08) found in  00-00-00-00-00-03
: FORWARD
DEBUG:forwarding.l2_firewall:installing flow for 00:00:00:00:00:08.3 -> 00:00:00
:00:00:01.1
DEBUG:forwarding.l2_firewall:Rule(00:00:00:00:00:01) found in  00-00-00-00-00-03
: FORWARD
DEBUG:forwarding.l2_firewall:installing flow for 00:00:00:00:00:01.1 -> 00:00:00
:00:00:08.3
DEBUG:forwarding.l2_firewall:Rule(00:00:00:00:00:01) found in  00-00-00-00-00-02
: FORWARD
DEBUG:forwarding.l2_firewall:installing flow for 00:00:00:00:00:01.1 -> 00:00:00
:00:00:08.3
DEBUG:forwarding.l2_firewall:Rule(00:00:00:00:00:01) found in  00-00-00-00-00-01
: FORWARD
DEBUG:forwarding.l2_firewall:installing flow for 00:00:00:00:00:01.1 -> 00:00:00
:00:00:08.2
DEBUG:forwarding.l2_firewall:Rule(00:00:00:00:00:01) found in  00-00-00-00-00-05
: FORWARD
DEBUG:forwarding.l2_firewall:installing flow for 00:00:00:00:00:01.3 -> 00:00:00
:00:00:08.2
DEBUG:forwarding.l2_firewall:Rule(00:00:00:00:00:01) found in  00-00-00-00-00-07
: FORWARD
DEBUG:forwarding.l2_firewall:installing flow for 00:00:00:00:00:01.3 -> 00:00:00
:00:00:08.2
DEBUG:forwarding.l2_firewall:Rule 00:00:00:00:00:02 not found in 00-00-00-00-00-
03 : DROP
DEBUG:forwarding.l2_firewall:Rule 00:00:00:00:00:02 not found in 00-00-00-00-00-
03 : DROP
DEBUG:forwarding.l2_firewall:Rule 00:00:00:00:00:02 not found in 00-00-00-00-00-
03 : DROP
DEBUG:forwarding.l2_firewall:Rule 00:00:00:00:00:02 not found in 00-00-00-00-00-
03 : DROP
DEBUG:forwarding.l2_firewall:Rule 00:00:00:00:00:02 not found in 00-00-00-00-00-
03 : DROP
DEBUG:forwarding.l2_firewall:Rule 00:00:00:00:00:02 not found in 00-00-00-00-00-
03 : DROP
DEBUG:forwarding.l2_firewall:Rule 00:00:00:00:00:02 not found in 00-00-00-00-00-
03 : DROP
DEBUG:forwarding.l2_firewall:Rule 00:00:00:00:00:02 not found in 00-00-00-00-00-
03 : DROP
DEBUG:forwarding.l2_firewall:Rule 00:00:00:00:00:02 not found in 00-00-00-00-00-
03 : DROP
DEBUG:forwarding.l2_firewall:Rule 00:00:00:00:00:02 not found in 00-00-00-00-00-
03 : DROP
DEBUG:forwarding.l2_firewall:Rule 00:00:00:00:00:02 not found in 00-00-00-00-00-
03 : DROP
```

Fig. 6 Debug messages for the pingall test demonstrating the flow rule installation and action on the packet based on the rules

Figure 8 explains the debug messages shown when the controller is started, the "Launch()" module is invoked and the firewall rules are added along with the startup of the controller. Rules are added in all the switches in the topology.

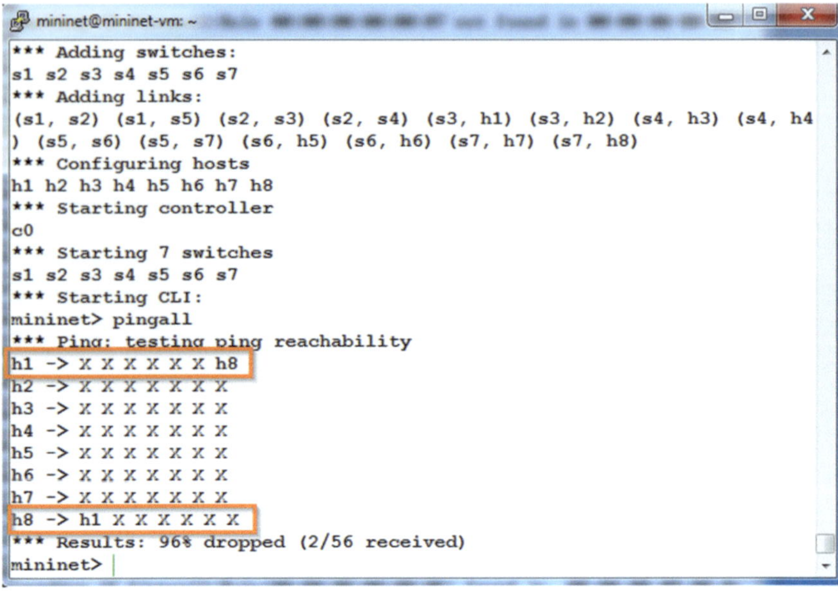

Fig. 7 Pingall test showing the successful modification of learning switch code to act as a firewall and allowing communication between h1 and h8 only

5 Conclusion and Future Scope

The algorithm and hence the L2 Learning switch code of the POX controller are successfully modified to insert rules and provide the firewall functionality for a bigger topology, i.e., a tree topology of depth 3. The modified code controls the traffic by adding rules. This implementation is not an intelligent firewall implementation. Further work would comprise adding the rules with source and destination MAC addresses, rather than the dpid of the switch and the source MAC Address, into a CSV file format by importing the file in the code, adding the rules to perform the firewall functionality. Further, the similar functionality for L3 Learning switch code is to be included by modifying the L3 Learning switch code.

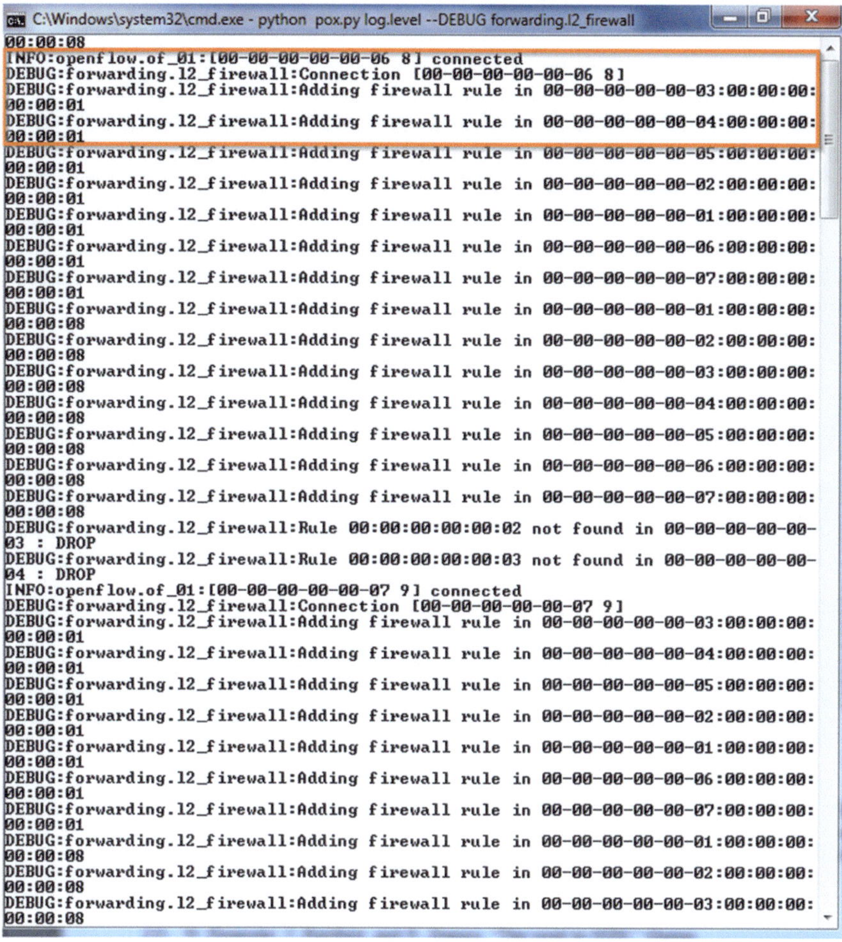

Fig. 8 The info message showing "Connection Up message" and the debug messages for "Addition of Firewall Rules"

Acknowledgements The authors would like to thank Mr. Santhosh Sundarasamy, Sr. Architect and Engineering Manager, Cloud Managed Security, Paladion Networks Pvt. Ltd., for his complete support and help in carrying out the experiment at Paladion Networks Pvt. Ltd as part of the internship program.

References

1. ONF (2013) SDN architecture overview version 1.0. Open Networking Foundation, December 2013
2. Nunes BAA, Mendonca M, Nguyen XN, Obraczka K, Turletti T (2014) A survey of software-defined networking: past, present, and future of programmable networks. Commun Surv Tuts 16(3):1617–16343 (IEEE)

3. Pfaff B, Davie B (ed) The Open vSwitch database management protocol. RFC 7047, December 2013
4. Heller B (2011) Openflow switch specification, version 1.0.0. Last accessed, Dec 2011. (Online). Available: www. openflowswitch.org/documents/openflow-spec-v1.0.0.pdf
5. Lantz B, Heller B, McKeown N (2010) A network in a laptop: rapid prototyping for software-defined networks. In: Proceedings of HotNets-IX. ACM, 2010, p 19
6. Stallings W (2013) Software defined networks and openflow, IPJ. (Online). Available: http://williamstallings.com/Papers/
7. Kaur S, Singh J, Ghumman NS Network programmability using POX controller. (Online) Available at: http://sbsstc.ac.in/icccs2014/Papers/Paper28.pdf
8. Feamster N (2014). SDN lectures 2014. (Online) Available at: https://class.coursera.org/sdn-001

Estimation Procedure of Improved High-Resolution DOA of Coherent Signal Source for Underwater Applications with Existing Techniques

Prashil M. Junghare, Cyril Prasanna Raj and T. Srinivas

Abstract Submerged target following in sea environment has pulled in extensive enthusiasm for both military and regular citizen applications. This paper displays the execution investigation of bearings of landing estimation procedures, subspace, and the non-subspace strategies. In this paper, investigating the Eigen-examination classification of high determination and super-determination calculations, presentation of depiction, correlation and the execution and determination investigations of these calculations are made. The examination is in light of direct exhibit receiving the wire and the count of the pseudo-spectra capacity of the estimation calculations. Customary MUSIC calculation breaks down the sign covariance network and afterward make the signs subspace acquired be orthogonal to the clamor subspace, which diminishes the impact of the commotion. Be that as it may, when the signs interim are little, customary enhanced MUSIC calculation has been not able to recognize the signs as the SNR diminishes. Another calculation is proposed utilizing SVD of the covariance lattice acquired. In this paper, different calculations are contrasted and every single accessible calculation. A ULA reception apparatus cluster setup is taken for both the calculations. Reproductions results demonstrate that proposed technique gives preferred execution over customary MUSIC calculation.

Keywords DOA · MUSIC · SNR · SVD · ULA

P. M. Junghare (✉)
Centre for Research and Development, PRIST University, Thanjavur, India
e-mail: prashiljunghare3@gmail.com

C. P. Raj
Electronics and Communication Department, MSEC, Bangalore, India
e-mail: cyrilprasanna007@gmail.com

T. Srinivas
Department of Photonics, Indian Institute of Science, Bangalore, India
e-mail: tsrinu@ece.iisc.ernet.in

A. Mishra et al. (eds.), *Silicon Photonics & High Performance Computing*,
Advances in Intelligent Systems and Computing 718,
https://doi.org/10.1007/978-981-10-7656-5_13

1 Introduction

The configuration and improvement of the shrewd exhibit reception apparatus is a standout among the most imperative examination subjects of cluster sign handling, which is firmly related remote correspondences, radar, radio stargazing, sonar, route, following of different questions, salvage, and other crisis help devices [1]. In late years, numerous critical exploration considerations have been pulled in the advancement Direction of Arrival estimation calculation, for example, Estimation of Signal Parameter through Rotational Invariance Technique (ESPRIT) algorithm [2], MUSIC calculation [3], adjusted MUSIC calculation. Sign preparing parts of savvy receiving wire frameworks has focused on the advancement of productive calculations for direction-of-arrival (DOA) estimation [4] and versatile pillar forming [5]. The late patterns of versatile pillar framing commute the advancement of computerized shaft shaping systems [6]. Instead of utilizing a solitary radio wire, an exhibit receiving wire framework with creative sign preparing can improve the determination of DOA estimation.

A cluster sensor framework has various sensors appropriated in space. This exhibit design gives spatial samplings of the got waveform. A sensor cluster has preferred execution over the single sensor in sign gathering and parameter estimation. Among the calculations of DOA estimation, as the super-determination spatial range estimation method, MUSIC calculation is a standout among the most established algorithms [7]. However, MUSIC calculation just can gauge applicable signs, when signs are a related or little distinction between the signs and SNR is low, the execution of the calculation reductions and even gets to be invalid. In this article, an adjusted MUSIC calculation is proposed utilizing the conjugate information. The solid consistency of the adjusted technique is built up. It is watched that the adjusted MUSIC works altogether superior to the common MUSIC at distinctive SNR as far as the mean squared lapse and for cognizant sources [8].

2 Mathematical Model and Preliminary Knowledge

MUSIC is an acronym which remains for Multiple Signal characterization [9]. It is high determination system in light of abusing the Eigen-structure of information covariance lattice. It is a straightforward, famous high determination, and productive Method. It guarantees to give fair-minded evaluations of the quantity of signs, the points of landing and the qualities of the waveforms [10, 11].

2.1 Mathematical Model

A uniform direct cluster (ULA) made out of N sensors and s narrowband signs of the diverse DOAs $[a(\theta_1)a(\theta_2)a(\theta_3)\ldots a(\theta_S)]$ was considered. At that point, a watched preview from the N exhibit components was demonstrated as

$$x(t) = A(\theta)m(t) + u(t), \tag{1}$$

where, $x(t)$ is the sign vectors at the cluster components yield, $m(t)$ is the sign vectors of the source, $u(t)$ is the noise vector at the array elements output, $A(\theta) = [a(\theta_1)a(\theta_2)a(\theta_3)\ldots a(\theta_S)]$ is the guiding framework, $a(\theta_S)$ is the cluster controlling vector relating to the DOA of the sign [12].

The cluster covariance grid T of the got signal vector in the forward heading can be composed as

$$T_{xx} = E\left[X(t)X(t)^H\right] = \frac{1}{L}\sum_{L}^{K}\left[X(t)X(t)^H\right], \tag{2}$$

where, K is the quantity of preview. MUSIC calculation, a piece graph of which can be found in Fig. 1, can be abridged in as takes after.

1. First, N tests from every collector channel must be gathered to shape $M \times N$ cluster. For reenactment purposes, this exhibit can be created by (2). Next, the covariance network T_{xx} must be estimated from received data.
2. Perform Eigenvalue decomposition on T_{xx}

$$T_{xx}E = E\wedge \tag{3}$$

where, $\wedge = \mathrm{diag}\{\lambda o, \lambda 1, \ldots \lambda_{N-1}\}$

$\lambda 0 \leq \lambda 1 \leq \ldots \lambda_{M-1}$ are the Eigenvalues and $E = [e_1 e_2 \ldots e_{N-1}]$ are the corresponding eigenvectors of T_{xx}.

3. Then, the DOAs of the numerous occurrence signs can be evaluated by finding the tops of the MUSIC range given by

$$P_{\mathrm{MUSIC}}(\theta) = 1/a(\theta)^H E_N E_N^H a(\theta), \tag{4}$$

where, $E_N = [e_{s+1}e_{s+2}\ldots e_{N-1}]$ is the subspace noise.

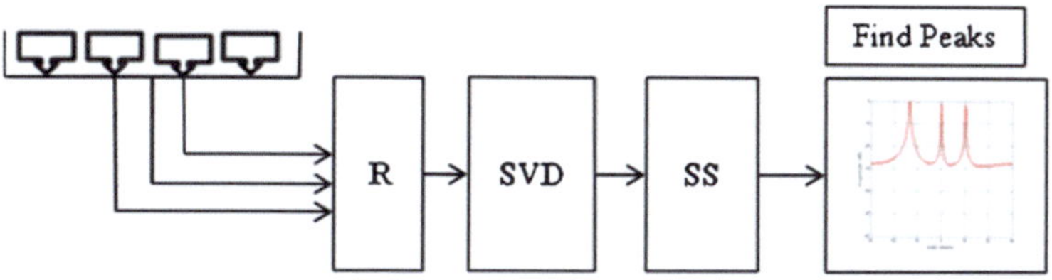

Fig. 1 Block diagram of MUSIC

Table 1 Under water parameters

Sl. no.	Parameters	Specifications
1	Water density	1000 kg/m^3
2	Speed of sound in water	1600 m/s
3	Water permittivity	80 F/m
4	Acceleration due to gravity	9.8 m/s^2

Now, after having obtained a MUSIC algorithm, we modify the steps so as to improve its performance [13].

4. The covariance matrix is decomposed using singular value decomposition given as

$$\mathrm{SVD}(T_{xx}) = \mathrm{USV}^H \quad (5)$$

A matrix TA can be calculated as

$$\mathrm{TA} = E_s \mathrm{EE}_s^H, \quad (6)$$

where, $E_S = [e_1 e_{s+2} \ldots e_{S-1}]$ is a signal subspace.

$$E = \mathrm{diagonal}(1/\mathrm{SS} - \mathrm{sigma} * I) \quad (7)$$

SS = diagonal (S_s) and SN = diagonal (S_N)

$$\mathrm{sigma} = \mathrm{trce}(S_N)/(N - D) \quad (8)$$

5. The new modified MUSIC algorithm is given by

$$P_{\mathrm{MUSIC}}(\theta) = a(\theta)^H * TA * a(\theta) / a(\theta)^H E_N E_N^H a(\theta) \quad (9)$$

The D biggest tops of the MUSIC range relate to the DOAs of the signs impinging on the cluster (Table 1).

3 Simulation Results

In this paper, the sound range of changed MUSIC is contrasted and the shaft filter calculation, most extreme entropy calculation, ESPRIT, MUSIC, root MUSIC, Modified MUSIC [14, 15]. It is cleared that from every single accessible calculation Modified MUSIC calculations gives the high determination and high exactness. This calculation works best notwithstanding when the sound sources are near one another.

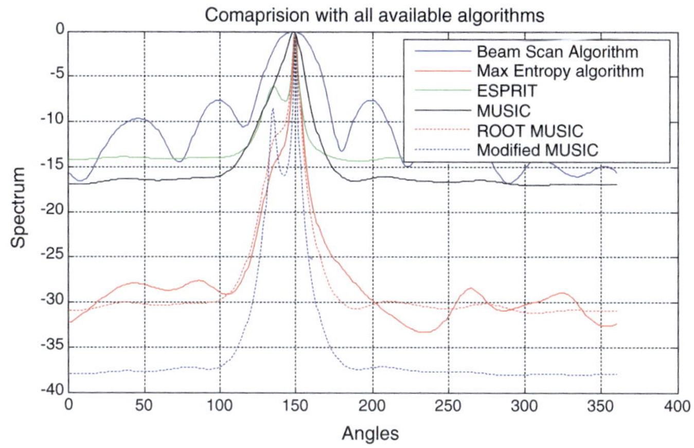

Fig. 2 Comparison of modified MUSIC with available algorithms

The Fig. 2 shows the comparison with the available algorithms. Root music algorithm fails when there is more noise. A MUSIC algorithm fails when the sound sources are very close to each other [16, 17].

4 Conclusion

Certain alterations are done in MUSIC calculation and by handling the covariance network of the exhibit yield flag, the proposed calculation for assessing DOA was produced. MUSIC calculation works fine for low-level commotion districts and its execution debases as the clamor level increments furthermore, neglect to separate signs which are close by. In that capacity, it cannot separate two signs isolated by AOA s of 4, under typical conditions.

The proposed calculation had the capacity recognize signals under a certain level of high commotion levels furthermore for near to sources. The adjusted calculation comes up short when the commotion level is expanded to a certain level. At such levels, it begins to identify clamor signals as the sought signs. So we have a tendency to get more tops in sign range diagram, making it hard to discover the genuine signs. Since we have utilized just ULA, the both broke down calculations give just azimuth edges.

Acknowledgements We, acknowledge Vision Group on Science and Technology (VGST), Govt. of Karnataka for providing research fund to carry out the work presented in this paper.

References

1. Chen J, Wu Y, Cao H, Wang H (2011) Fast algorithm for DOA estimation with partial covariance matrix and without eigen decomposition. J Sig Inf Process 266–269
2. El-Barbary KA, Mohamed TS, Melad MS (2013) High resolution direction of arrival estimation (coherent signal source DOA estimation). Int J Eng Res Appl (IJERA) 3(1):132–139. ISSN:2248-9622
3. Zhang X, Xu L, Xu L, Xu D (2010) Direction of Departure (DOD) and Direction of Arrival (DOA) estimation in MIMO radar with reduced-dimension MUSIC. IEEE Commun Lett 14 (12):1161–1163
4. Zhao Q, Ai Z (2012) An iterative MUSIC algorithm research based on the DOA estimation. Int Conf Biol Biomed Sci 1–5
5. Chetan R, Dongarsane, Jadhav AN (2011) Simulation study on DOA estimation using MUSIC algorithm. Int J Technol Eng Syst (IJTES) 2
6. Omar MMM, Mohamed DAE, Haikel SM (2011) The mutual coupling effect on the MUSIC algorithm for direction of arrival estimation. Int J Comput Appl (0975–8887) 35(4)
7. Zhang Y, Obeidat BA, Amin MG (2006) Spatial polarimetric time-frequency distributions for direction-of-arrival estimations. IEEE Trans Sig Process 54
8. Zhou QC, Gao HT, Wang F (2012) A high resolution doa estimating method without estimating the number of sources. Prog Electromagnet Res C 25:233–247
9. Santhosh S, Sharma K (2013) A review on multiple emitter location and signal parameter estimation. Int J Eng Res 2(3):237-242. ISSN:2319-6890
10. Mohamed DA, Haikel SM, Omar MM (2011) The mutual coupling effect on the MUSIC. Int J Comput Appl 35(4):36–40
11. Lee KH (2013) Improve method on DOA estimation accuracy of signal subspace sensitivity based on mutual coupling matrix. Int J Smart Home 7(3):1–10
12. Yu Y, Member, Lui H-S, Niow CH, Hui HT (2011) Improved DOA estimations using the receiving mutual impedances for mutual coupling compensation: an experimental study. IEEE Trans Wirel Commun 10(7):2228–2233
13. Li J, Zhang X, Cao R, Zhou M (2013) Reduced-dimension MUSIC for angle and array gain-phase error estimation in bistatic MIMO radar. IEEE Commun Lett 17(3):443–447
14. Shahedul Amin Md, Riyasat Azim Md, Rahman SP, Ferdous Habib Md, Ashraful Hoque Md (2010) Estimation of Direction of Arrival (DOA) using real-time array signal processing and performance analysis. Int J Comput Sci Netw Secur 10(7):43–47
15. McCloud ML, Scharf LL (2002) A new subspace identification algorithm for high resolution DOA estimation. IEEE Trans Antennas Propag 50(10):1382–1390
16. Kundu D (1996) Modified MUSIC algorithm for estimating DOA of signals. Sig Process 48:85–90
17. Evans TE (1981) High resolution angular spectrum estimation technique for terrain scattering analysis and angle of arrival estimation. In: Proceedings of 1st ASSP workshop spectral estimation, communications research laboratory, McMaster Univ., Hamilton, Ont., Canada, pp 134–139

Finite Element Analysis of Fiber Optic Concentric Composite Mandrel Hydrophone for Underwater Condition

Prashil M. Junghare, Cyril Prasanna Raj and T. Srinivas

Abstract An Interferometric fiber optic hydrophone is designed in this work with a composite concentric structure. The structure is made of different layers having a variable material and structural properties. The mandrel is designed to withstand a natural frequency ranges from 0.2 to 2.5 kHz. The objective of the work is to design the mandrel which is placed at a distance ranging from 20 to 200 m underwater with varying boundary conditions. Boundary conditions specified are innermost layer of the mandrel is fixed and pressure is applied to the outermost layer of the mandrel. The design is feasible with two optic fiber layers which are wound over the center of the length of the mandrel. Preprocessing of design is made using Hyper Mesh; analysis is performed in ABAQUS 6.10 CAE tool and visualization of results in hyper view. Whenever the pressure is applied to the mandrel, the phase change of light happens which can be to calculate sensitivity mathematically.

Keywords Interferometric · Concentric structure · Hyper mesh
Hyper view · ABAQUS CAE · Preprocessing

P. M. Junghare (✉)
Centre for Research & Development, PRIST University, Thanjavur, India
e-mail: prashiljunghare3@gmail.com

C. P. Raj
Electronics & Communication Department, MSEC, Bangalore, India
e-mail: cyrilprasanna007@gmail.com

T. Srinivas
Department of Photonics, Indian Institute of Science, Bangalore, India
e-mail: tsrinu@ece.iisc.ernet.in

A. Mishra et al. (eds.), *Silicon Photonics & High Performance Computing*,
Advances in Intelligent Systems and Computing 718,
https://doi.org/10.1007/978-981-10-7656-5_14

1 Introduction

In present days, sensor technology has become a vast concept and is grabbing the world's attention toward it. A number of innovative terminologies related to the field of sensors is evolving day to day. The updating of terminologies in sensor technology is leading to modify the existing design; many dissertation works are carried under reputed research and development centers, universities. These canters are providing a platform for research scholars in the field of sensor technology. Flexible designs are evolving with the technology improvement.

Interferometry with a medium of electromagnetic waves is considered and superimposed for extracting the information about waves. The optical fiber is a device that uses the effect of interference [1]. Here an input beam is split into two with the use of beam splitter and some of these beams are exposed to external influences such as length change or refractive index change in the transparent medium [2]. The beams are recombined by a beam coupler. In the design two mandrels are used, one is considered as reference mandrel which provides results under ideal conditions and another is a sensing mandrel, whose output is to be measured. Reference mandrel is designed such that, it does not respond to any of the environmental changes. But sensing mandrel will respond to changes that are occurred naturally. Sensing mandrel is designed to be placed in underwater; hence, it is called as hydrophone [3] (Fig. 1).

The purpose of placing sensing mandrel in underwater is to detect the pressure variations known as acoustic pressure [4]. Different materials are considered in designing hydrophone they are Nylon, Aluminum, Polystyrene, Optic fiber, and Polyurethane. The design has to be feasible to suspend the hydrophone under the water at a depth of 200 m. Depending on requirements the depth may be varied [5].

The final output required from the hydrophone is sensitivity which is expressed in terms of decibels. To get the optimum sensitivity dimensional parameters are varied; such as effective length, diameter, and thickness. Finite element analysis [6] of hydrophone is performed to know the preferable design. Light form source is passed through an optic fiber that is wound in between Polystyrene and Polyurethane layers. While light is propagating through optic fiber the pressure load is applied to the sensor, change in phase is obtained. Change in the phase of both

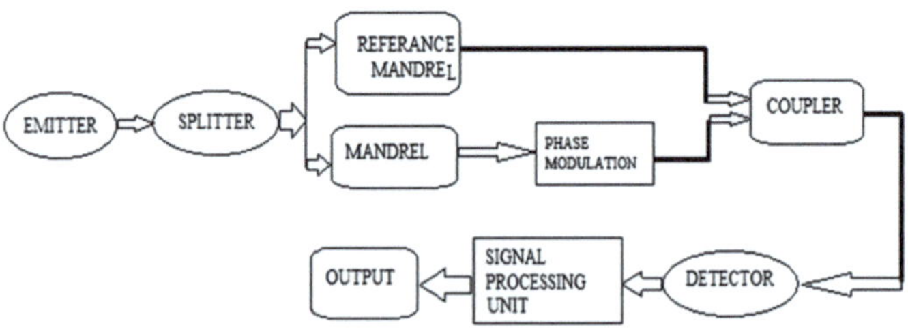

Fig. 1 Block diagram of phase change detection process

reference mandrel and sensing mandrel are coupled using a coupler and passed to the detector, where the output is converted into electrical signal and is processed in the signal processing unit [7]. Here, phase change is detected by comparing the output of both reference mandrel and sensing mandrel.

2 Mach–Zehnder Interferometric Principle

Mach–Zehnder interferometer [8] is shown in Fig. 2 Configuration of this kind represents a typical transmissive type fiber optic hydrophone. It can be used to build a transmission-type sensor array. Here, laser light is split into two beams by the first fiber beam splitter, one is entering sensing arm and the other is entering reference arm. In a single sensor case, the sensor head in the sensing arm is placed in the sensing environment. The reference arm provides the phase under the ideal condition and it can stay with all other components of the system at the "dry" end. Near the output end of the second fiber coupler, the two beams are sent to a photodetector in a combined manner, which produces an electrical signal resulting from the interference of the signals of sensing and reference beams at the receiver circuit.

3 Basics of Finite Element Analysis

Eventually, each and every phenomenon in nature related to biological, geological or mathematical can be defined with aid of laws of physics. Historical background related aspects are required for the mathematical formulation of the physical process. Formulation results in form of mathematical statements [9], often differential equations relating quantities of interest in understanding and design of the physical process. Assumptions related to proceedings of work carried are needed for the development of the physical process. Derivation of governing equation for a complex problem is not so difficult [10]. Finding their solution by exact analysis is a time-consuming job. The value of desired unknown quantities at any location in a

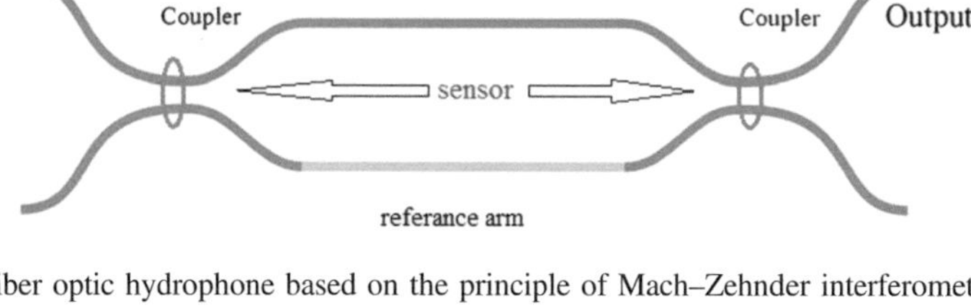

Fig. 2 Fiber optic hydrophone based on the principle of Mach–Zehnder interferometer

body can be found using analytical solution which is in the form of mathematical expression. For idealized and simplified situations analytical solution can be easily obtained.

4 Modeling and Design

1. Preparation of CAD model using CATIA V5 R20.
2. Meshing the model by using Altair's HYPERMESH 12.
3. DECK prepared by using Altair's HYPERMESH 12.
4. Analysis is performed in ABAQUS CAE 6.10 solver.
5. Required output such as strain values can be viewed either using Altair's HYPERVIEW or ABAQUS CAE 6.10.

The optic fiber used is made of silica glass having diameter 125 micrometers which is wound around the polystyrene layer of the hydrophone. Material, total number of elements used, thickness of elements, and area occupied by these elements is shown in Table 1 (Figs. 3 and 4).

Table 1 Material, Elements, Area, and Thickness of layers

Sl/no	Material	Number of elements	Area occupied (mm^2)	Thickness (mm)
1	Nylon	744	4430.123	1.5
2	Aluminum	6138	10102.437	10
3	Polystyrene	26784	26635.680	20
4	Optic fiber	3600	13901.519	0.25
5	Polyurethane	9000	20966.297	10
Total assembly		46266	31117.624	

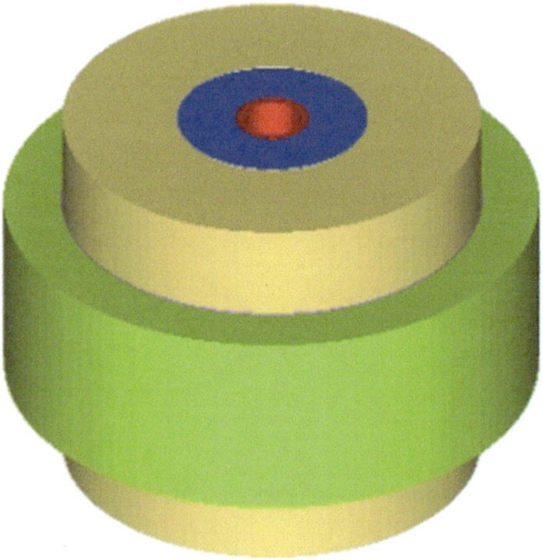

Fig. 3 CAD assembly of hydrophone 3

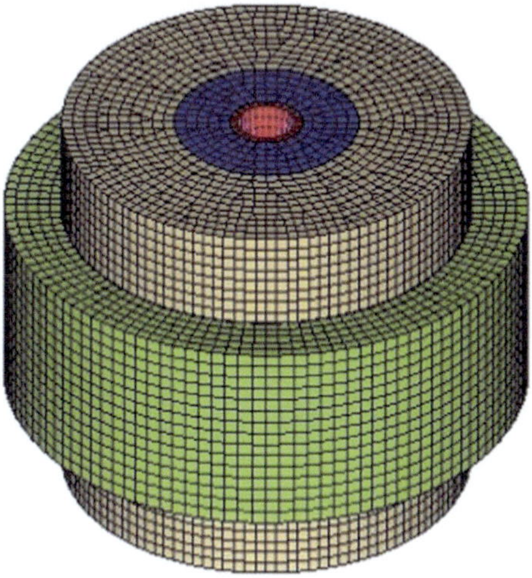

Fig. 4 Meshed model of hydrophone

5 Results and Discussion

5.1 *Static Analysis of Sensing Mandrel*

Static analysis is performed to know the variations in the sensing mandrel when it at rest with constant pressure loads on the outermost part of sensing mandrel. When the mandrel is placed in underwater condition at a depth of 200 m, it is assumed that a constant hydraulic pressure of 2 Mpa is applied. Results considered in this model are displacement model, axial strain model, radial strain model, and stresses in the model.

$$P = \rho g h,$$

where,

P density of medium = 1000 kg/m^2
g Gravitational force = 9.81 N
h Depth, = 200 m

$$P = 1000 \times 9.81 \times 200 = 1{,}962{,}000 = 1.962 \text{ Mpa}$$

$$P = \sim \mathbf{2\,Mpa}$$

Sensitivity of hydrophone can be found using the following relations, where φ is the phase of light propagating, P is externally applied acoustic pressure, also S_r = 1 rad/μPa, n is the refractive index of fiber core and ε_r **is** radial strain acting on the surface, ε_z is the axial strain resulting the externally applied acoustic pressure.

$$\Delta\varphi/\varphi = \varepsilon_r + \varepsilon_z - n^2/2 \cdot [(p_{11} + p_{12}) \cdot \varepsilon_r + p_{12}\,\varepsilon_z],$$

where

ε_r = 1.648×10^{-3},
ε_z = 5.954×10^{-4},
P_{11} = 0.121,
P_{12} = 0.27,

$$S_m = \Delta\varphi/p$$

Sensitivity of hydrophone is given by, $S = 20 \log (S_m/S_r)$, where $S_r = 1\mu$/rad

$$S = 20 \log (S_m).$$

5.2 *Influence of Geometric Properties on Sensitivity of Hydrophone*

Geometry of hydrophone affects the sensitivity with the following parameters

- Inner to outer diameter ratio
- Outer diameter
- Thickness of foaming layer
- Optic fiber length

The below shown figure gives the brief idea that how sensitivity is affected. Here, when the hallow diameter is for hydrophone is taken as zero or no hallow diameter sensitivity was seen comparatively less. Whereas with a hallow diameter sensitivity was seen to be improved. When the inner to outer diameter ratio was varied with the variable length of optic fiber increase in sensitivity was achieved. Variation is shown in the below figure. Constant improvement in sensitivity was achieved till inner to outer diameter ratio of 65%, above the sensitivity improvement is but this improvement is because of variable length of the optic fiber.

Fiber optic length is directly in relation with the phase change of light (φ), so that change in optic fiber length results in a change in sensitivity. The figure shown below indicates that, as the length of optic fiber increases sensitivity increases. But it is not possible to consider the higher length of the optic fiber, because with an increase in the length of the optic fiber effective length of mandrel increases and which affects the light propagation through optic fiber (Figs. 5, 6, 7 and 8).

Outer diameter is also a parameter that affects the sensitivity of hydrophone. Without considering inner (hallow) diameter sensitivity produced is comparatively less. But when inner hallow diameter considered increase in sensitivity is achieved. Even with the increase in outer diameter, length of optic fiber wound around the outer diameter increases which in turn increase sensitivity.

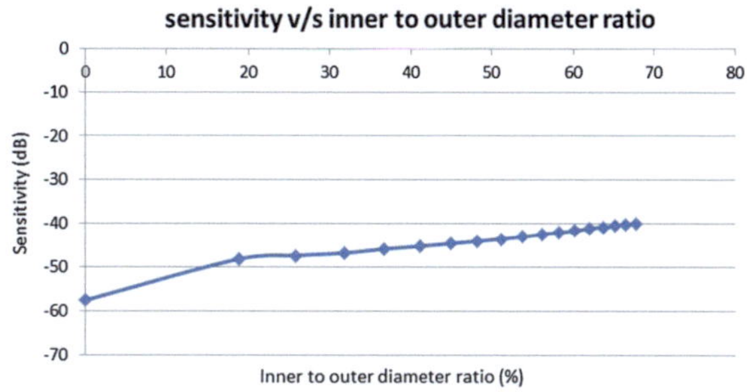

Fig. 5 Sensitivity versus inner to outer diameter ratio

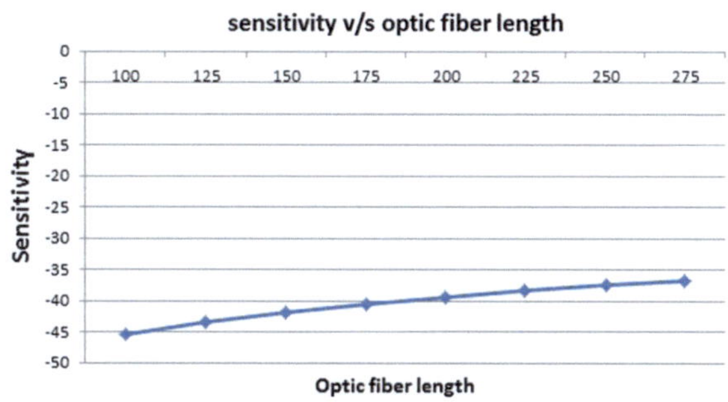

Fig. 6 Length of optic fiber versus sensitivity

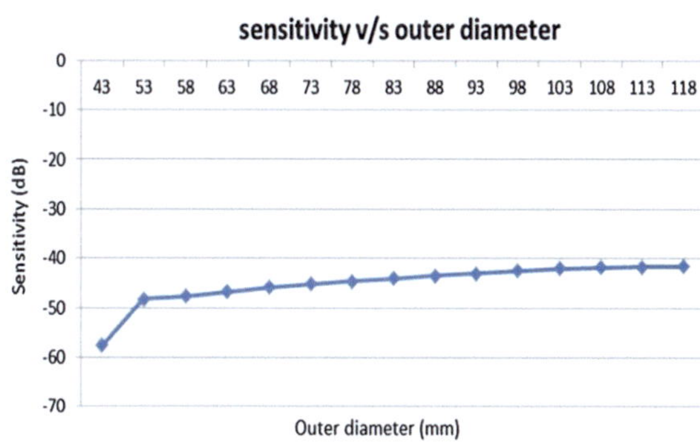

Fig. 7 Sensitivity versus outer diameter

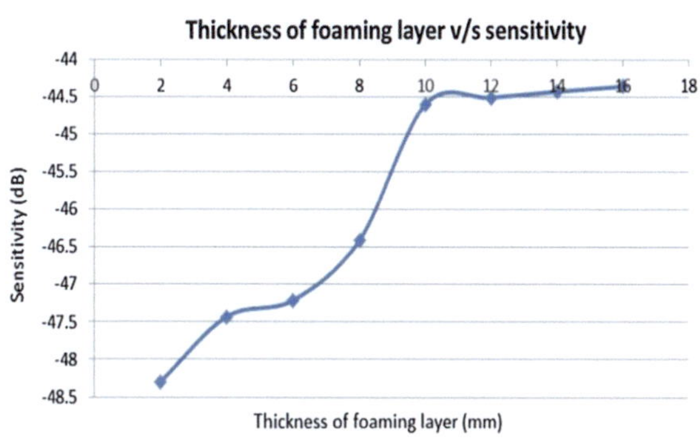

Fig. 8 Sensitivity versus thickness of foaming layer

Similarly, when the polystyrene material is used that allows the optic fiber to flexibly move (stretch), and thus, it can sense the acoustical pressure applied. As the thickness of foaming layer increases sensitivity also increases till it reaches (thickness of aluminum to thickness of polystyrene) 1:1 ratio. Even after that sensitivity increases slightly this is because of increase in outer diameter of sensing mandrel.

Also material properties have an influence on the sensitivity of hydrophone.

- Young's modulus,
- Poisson's ratio,
- Type of material (ductile or brittle).

Material properties also affect the sensitivity of hydrophone. Materials which are flexible in nature will tend to behave as ductile, which means they deform elastically. The materials which are having lowest young's modulus can stretch with small loads. Hence, the layer between optic fiber and aluminum must have lowest young's modulus. So many materials are there which have lowest young's modulus; among them, some of them are used to find sensitivity. Polystyrene is one which gives more sensitivity about −43.56 dB. Poisson's of individual material will not affect the sensitivity, but there will be slight variation about 1–2 dB.

Sometimes variation in sensitivity is too small that can be neglected. If the material used is ductile material then strain is produced, whereas when the brittle material is used more acoustical pressure load is required to produce strain (Figs. 9 and 10).

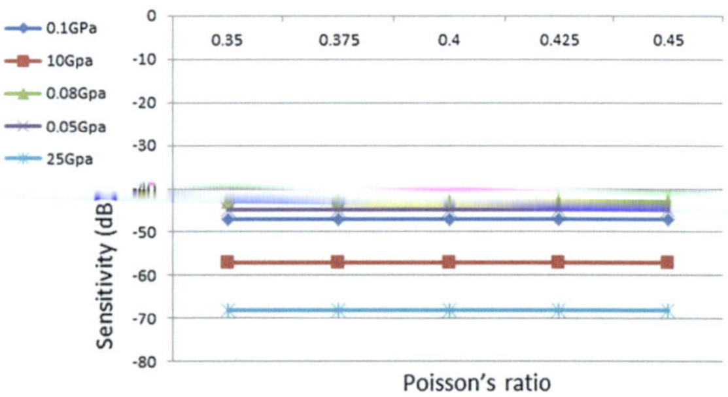

Fig. 9 Sensitivity versus Poisson's ratio

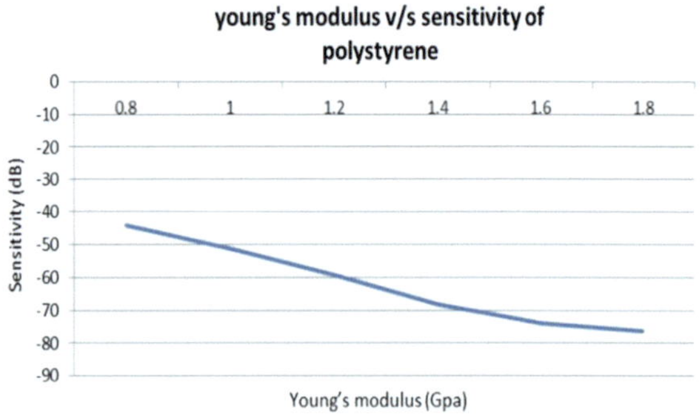

Fig. 10 Young's modulus versus sensitivity

6 Conclusion

This project outlines the finite element analysis of Mach–Zehnder fiber optic Hydrophone. Design indicates the structure of concentric composite mandrel. The foaming layer of polystyrene is used with base material as Aluminum. The hydrophone is designed to have fundamental natural frequency over 2.5 kHz was achieved from the subsequent analytical test. Better sensitivity was achieved when it is underwater at a depth of 200 m. While designing hydrophone the parameters considered are material properties of Polystyrene layer, Aluminum layer, Optic fiber, and Polyurethane layers along with the geometry of hydrophone. By viewing and analyzing the results, it is found that sensitivity has drastically improved. During the design properties of the material used showed that, Polystyrene (foaming layer) of the mandrel is having lowest young's modulus with respective Poisson's ratio.

In the design, by varying effective length and thickness of optic fiber considerable change in sensitivity was obtained. The mandrel is designed for resonance frequency about 15 kHz with the optimal design of concentric composite mandrel, the sensitivity of −43.27 dB is obtained with respect to $S_r = 1$ rad/μPa for the applied pressure load of 2 Mpa. The results shown have 20 dB increased sensitivity over conventional hollow cylindrical mandrel type hydrophone.

Acknowledgements We, acknowledge Vision Group on Science & Technology (VGST), Govt. of Karnataka for providing research fund to carry out the work presented in this paper.

References

1. Africk S, Burton T, Jameson P, Ordubadi A (1981) Design studies for fiber optic hydrophones. Report No. 4658. Bolt, Beranek & Newman, Inc., Cambridge, Mass
2. Hughes R, Jarzynski J (1980) Static pressure sensitivity amplification in interferometric fiber optic hydrophones. Appl Optics 19(1):1
3. Im J-i, Roh Y (1999) A finite element analysis of an Interferometric optical fiber hydrophone. Am J Eng Res
4. Hocker GB (1979) Fiber optic sensing of pressure and temperature. Appl Opt 18(9):1445
5. Wang Z, Hu Y, Meng Z, Ni M, Luo H (2010) College of Photoelectric Science and Engineering, National University of Defense Technology, 410073, P. R. China
6. Stockbridge N (2007) Fiber optic hydrophones. Winfrith Technology Centre, Winfrith Newburgh, Dorchester, vol 6619, 661907-2
7. Shahrieel M, Aras M, Mazni M, Hairi H, Jamaluddin MH (2010) Development of hydrophone sensor system for autonomous underwater vehicle application. In: Second international conference on engineering and ICT
8. Mach-Zehnder Interferometer sensor for acoustic detection with optimal performance. IOSR J Electron Commun Eng (IOSRJECE) 2(5):29–33, ISSN: 2278–2834 (2012)
9. Arshad MR (2009) Indian J Mar Sci 38:267–273
10. Lagakos AN, Hickman TR, Ehrenfeuchter P, Bucaro JA (1991) Planar flexible fiber-optic acoustic sensors. In: Proceedings of the 34th midwest symposium on circuits and systems, vol 56, pp 666–671

A Simulation Study of Design Parameter for Quantum Dot-Based Solar Cells

Ashwini A. Metri, T. S. Rani and Preeta Sharan

Abstract Solar cells are constructed with the incorporation of Quantum dot layers in it. This technology of using Quantum dots in the solar cells now has become an emerging area of research in the nanotechnology field. Developing of Quantum dot-based solar cells with the use of different material layers to study the electrical and optical characteristics of a solar cell which influences the performance. Crosslight-APSYS software tool has been used to design the quantum dot solar cells using ZnO/CdZnO as the photosensitive layer. Bandgap energy, concentration of holes and electrons, current trap occupancy graphs, are obtained form the simulations.

Keywords Bandgap · Multi-junction · Miniband

1 Introduction

Solar cell design that uses quantum dots as absorbing photovoltaic material is called the Quantum dot-based solar cells. The variation in the thickness of this layer affects the bandgap energy which in turn affects the design parameters [1, 2]. Most commonly used materials are copper indium gallium selenide (CIGS) or CdTe which are huge in size. Most popularly used material is silicon. When these layers are replaced by quantum dot solar cells having bandgaps that are tunable over a widespread range of energy levels just by varying the size of these dots. As a common practice, the bandgap of bulk materials is determined by the type of material(s) chosen. Due to this property quantum dots are more agreeable for

A. A. Metri (✉) · T. S. Rani · P. Sharan
Department of ECE, The Oxford College of Engineering, Bengaluru, India
e-mail: ash.metri@gmail.com

T. S. Rani
e-mail: ranianits@gmail.com

P. Sharan
e-mail: sharanpreeta@gmail.com

A. Mishra et al. (eds.), *Silicon Photonics & High Performance Computing*,
Advances in Intelligent Systems and Computing 718,
https://doi.org/10.1007/978-981-10-7656-5_15

multiple junction solar cells, where a range of materials are used to enhance the efficiency by laying various portions of the solar spectrum. The popularity of quantum dots for use of solar cells is due to its propensity to tune the bandgap.

Assimilation of QWs or QDs into the active region of a photovoltaic device permits a grade of control over the electrical and optical properties of the complete system. This property is an outcome of quantum confinement effects which appear due to the discontinuities in the probable energy band edges of the nanoscale layers of alternating semiconductor material and acutely alter carrier energy levels and their dynamics. Hence, by an apt choice of individual layer thickness and compositions, the nanostructures can be tuned to transform a specific color of light into electricity more efficiently.

Quantum dots are more preferred for a further more reason of being cost-effective with respect various reasons such as first that they can be produced economically and their ability yield more energy. Second, mass production of quantum dots through high-throughput roll-to-roll manufacturing helps to lower the cost further more. Third, additional electricity can be derived for every photon of light, through multiple electron generations. As on date in the market, silicon solar cells are most efficient, electricity converted is less than 20% of the light that hits them, and where the theoretical maximum efficiency of these cells is around 31%.

2 Structure of Quantum Dot Solar Cell

The structure in Fig. 1 is an example of Quantum dot Solar cell which is made up of a material Zno/CdZno is constructed and simulated using Crosslight apsys software. Using an effective miniband model within the framework of drift-diffusion theory, advantages of QW/QD solar cells can be demonstrated via simulation (Fig. 2).

Fig. 1 Model of quantum dot solar cell

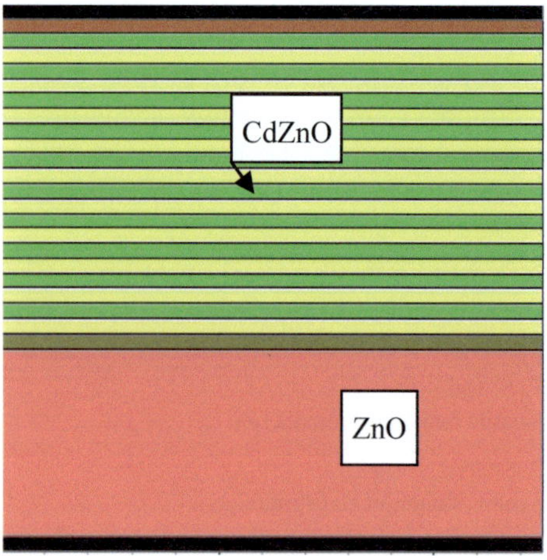

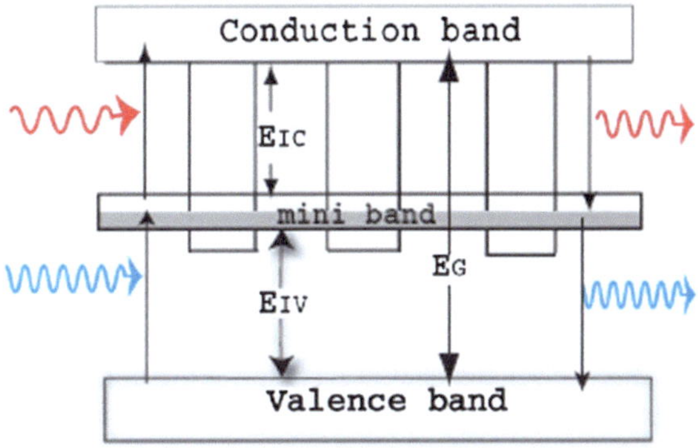

Fig. 2 Band diagram

With different energies of the effectual miniband, hot, and cold carrier miniband can be efficiently simulated. Expanded raptness of wavelength variety for a given material. Miniband transports of either hot carrier or cold carrier supposedly enhance carrier lifetime. Hot carrier miniband transport to render higher open-circuit voltage. Cold carrier miniband is used to transport and improvise short-circuit current. QD solar cell is made up of Zno/CdZno. This will be done by putting details for each layer like material of composition, height, thickness, doping type of either *n*-type or *p*-type, concentration, etc. These are the parameters available in the Crosslight-APSYS tool.

3 Method of Proposition

Quantum dot Solar cells are constructed and simulated using crosslight APSYS software. APSYS, Advanced physical models of semiconductor devices does 2D or 3D finite element analysis of electrical, thermal, and optical characteristics of compound semiconductor devices, in consideration to silicon as a special case. We have designed a solar cell device using quantum dots by the use of different materials, such as Zno/CdZno, acting as a bulk and active material and efficiency can be increased up to 45% in the multijunction solar cells. Optical and electrical properties can be studied by obtaining graphs of Parameters like electron concentration, hole concentration, mobility, total current, internal quantum efficiency, spectrum rate, bandgap energy, I–V characteristics, etc. can be inferred from the graphs.

By changing of material composition which is more efficient and the inclusion of multijunction Quantum dot layers the efficiency can be increased up to 31–45% when compared with normal solar cells. Since the equilibrium pressure of zinc is significantly higher than that of cadmium, zinc tends to form droplets of quantum dots. The fundamental physical quantities, such as band diagram, optical absorption, and generation for the modeled TJ (tandem junction or multijunction) cell, and external quantum efficiency for distinctive subcell junctions are procured.

4 Results and Discussion

Simulation results have been derived for quantum dot solar cells for various parameters. The spectral behavior for various parameters of quantum dot solar cells has been discussed below. The design parameters obtained are bandgap energy, electron and hole concentrations, field potential, conductivity, and trap occupancy. The input parameters given for the design layers are *n*-type concentration is 2.5 m, *p*-type concentration is 1.64 m, thickness of the solar cells 2.5 μm.

4.1 Simulation Procedure

The incident light required for testing the structure is applied by *.sol file. In case of these experimental results file chosen to use is AM15 solar spectrum as the optical pumping source: provided in the solar.am15. The light input is confined to simulate the shadowing effect of the contacts to the center of the device. Boundary conditions are important in addition to the input light. The other important point to be considered is that it should have a broad range of wavelengths in solar cell simulations. Therefore, a single value of the complex refractive index *(n, k)* is inappropriate. Hence to reason the barrier-depletion or voltage-drop effect, simulation using drift-diffusion model only has to be done on the highly *n*-doped junction of ZnO/CdZnO with electrons flowing from smaller bandgap ZnO. Different simulated graphs are obtained by feeding the input values for it.

From Fig. 3 shows variations of bandgap energy with thickness (distance) for ZnO Solar cells. The red and purple line shows energy levels for electron and holes, respectively. Mobile carriers are depleted due to the Peak of the potential barrier while higher electric field layer is generated by the ionized dopants. High internal field creates a steep potential profile which pushes the barrier higher.

The behavioral graph in Fig. 4 shown for depletion region in case of multi-junction design for the carrier concentration of holes and electrons in the multi-layered quantum dot solar cell.

By considering the parameters, such as *n*-type, *p*-type concentrations, for each individual layer and their layer thickness following spectral graph is obtained in Fig. 5. It shows the combined current for both electrons and holes concentration.

The performance of solar cell device can be explained by its I–V characteristics shown in Fig. 6. It represents the conductivity of the solar cell, consistency in current can be observed for a range of voltage. The sharp drop in current is

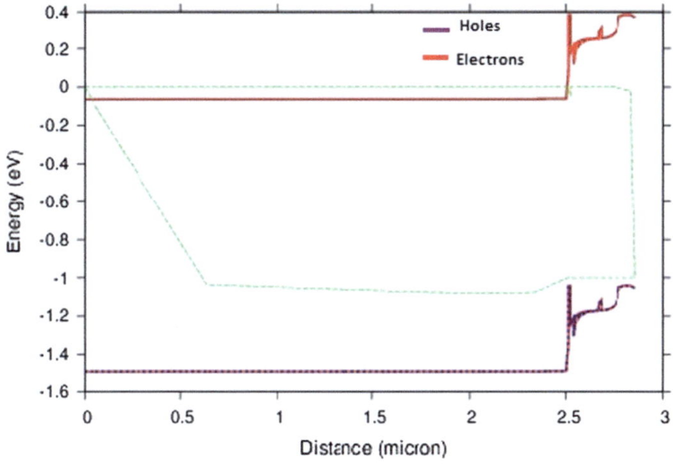

Fig. 3 Bandgap energy verses distance moved along the thickness of ZnO Solar Cell

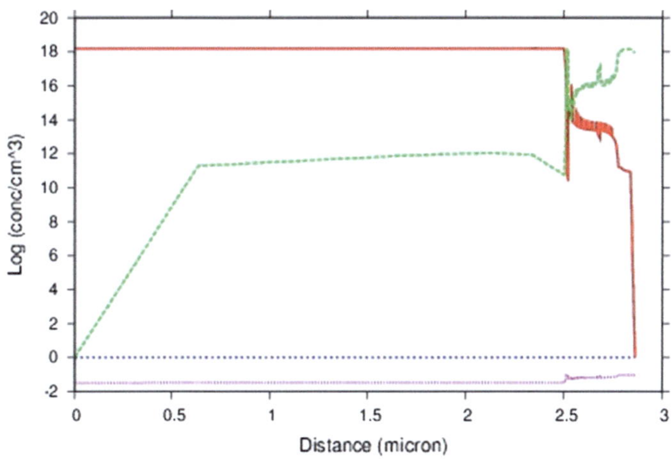

Fig. 4 Graph of total concentration indicating both electron and hole concentration in the bandgap region of solar cell

observed at a thickness of 2.5 μm, which is the maximum thickness of the solar cell designed. This, in turn, shows their negligible loss in the current.

When a voltage bias is applied from the side of ZnO the device is reverse biased and carriers are unable to flow into the depletion layer with which the applied field adds to the internal field to create a wider depletion layer. The carrier's are then made go over the barrier by the influence of quantum tunneling effect [3]. Introduction of quantum dot layers in between ZnO and CdZnO regions in the

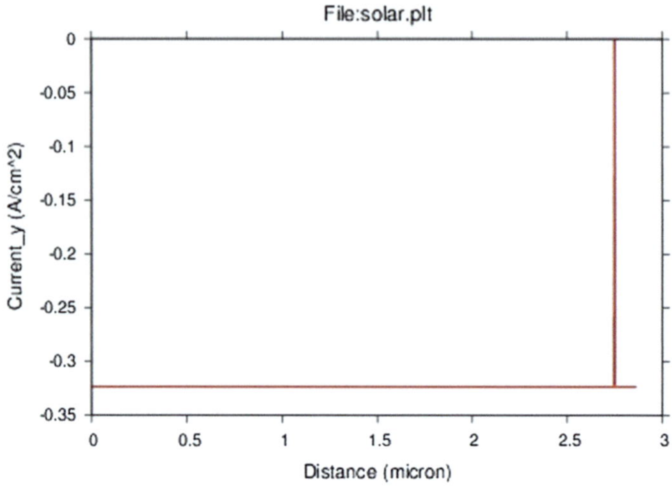

Fig. 5 Graph of total current which is a combination of both electron and hole current on *y*-axis

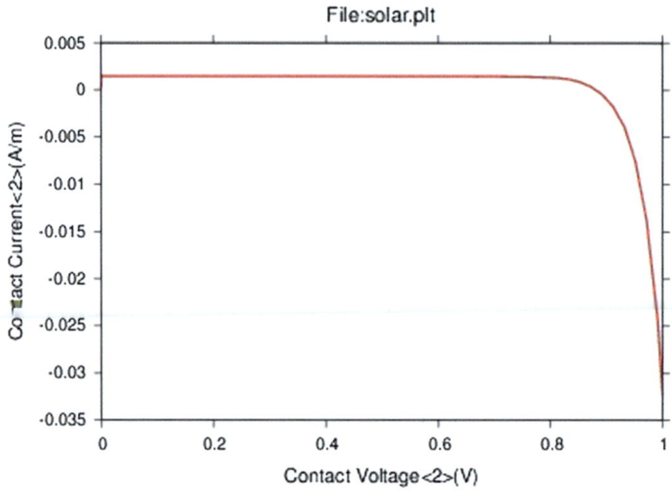

Fig. 6 I–V characteristics of ZnO/CdZnO

multilayered quantum dot solar cell design has brought a change in the field which can be observed in Fig. 4 [4, 5]. This change, in turn, helps in increasing efficiencies up to 45% (Fig. 7).

Trap occupancy represents the rate of mobility of charge carriers which can be observed in Fig. 8 [6]. It shows that the rate of absorption until 2.5 μm width of a device is maximum and further fluctuates and deteriorates at a boundary condition [7, 8].

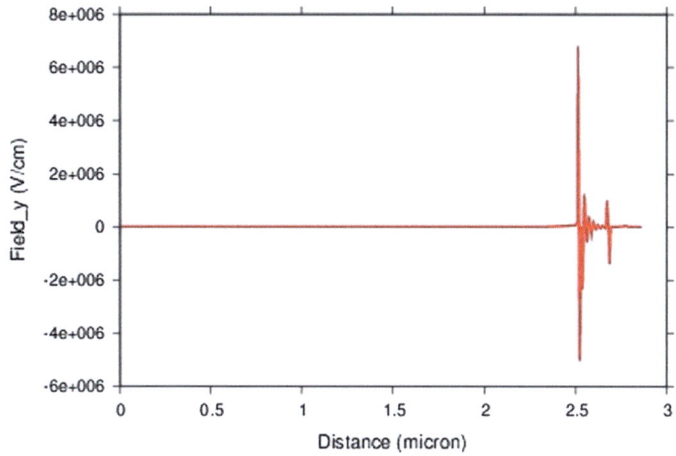

Fig. 7 Variation of electric field along *y*-axis of a device

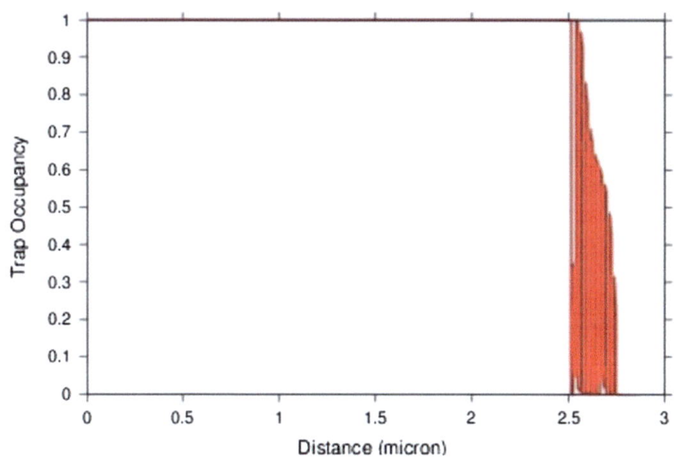

Fig. 8 Graph of trap occupancy

5 Conclusion

A pin structure of ZnO/CdZnO solar cell is designed with tool Crosslight-APSYS. Layers of stack Quantum dot layers are embedded between the ZnO/CdZnO. From the simulations obtained the optical and electrical characteristics and performance behavior has been successfully inferred. This method gives a selection of material for a solar cell in order to derive good performance. This methodology can be tested further on different materials at different bandgap energy levels. Calculation of efficiency for which material the value will be maximum can also be obtained with the optical characteristics designed above.

References

1. Xiao YG, Uehara K, Lestrade M, Li ZQ, Li ZS (2009) Modeling of si-based thin film triple-junction solar cells. In: IEEE Photovolatic specialists conference (PVSC), pp 002154–002158
2. Xiao YG, Li ZQ, Lestrade M, Li ZS (2011) Modeling Of Cdznte/Cdte/Si triple junction solar cells. In: IEEE-Photovolatic specialists Conference (PVSC), 2011, pp 01386–001389, 0160–8371
3. Driscoll K, Hubbard S (2011) Modeling the optical and electrical response of nanostructured III–V solar cells. In: IEEE Photovolatic specialists Conference (PVSC), 2011, pp. 002985–002989
4. Mammar H, Benmansour A, Bouzaki M (2014) Improvement of the photovoltaic conversion efficiency using nanostructuring in intermediate- band photovoltaic solar cells. In: IEEE-Dielectric materials for Photovoltaic systems (NAWDMPV), pp. 1–7
5. www.crosslight.com
6. http://www.gizmag.com/quantum-dot-solar-cells/32478/
7. http://www.scientificamerican.com/article/quantum-dots-and-more-use/
8. http://www.pveducation.org/pvcdrom/pn-junction/band-gap

MIX
Papier aus verantwortungsvollen Quellen
Paper from responsible sources
FSC® C105338

If you have any concerns about our products,
you can contact us on
ProductSafety@springernature.com

In case Publisher is established outside the EU,
the EU authorized representative is:
Springer Nature Customer Service Center GmbH
Europaplatz 3, 69115 Heidelberg, Germany

Printed by Libri Plureos GmbH
in Hamburg, Germany